An accounting approach to create an environmentally sustainable company

Monograph Series MEYER STIFTUNG

Edited by
Prof. Dr. Claus Meyer

Volume 18

Tanja Schillinger

An accounting approach to create an environmentally sustainable company

Selection and Definition of Environmental Indicators with special reference to Suppliers in Developing Countries

Verlag Wissenschaft & Praxis

Bibliographic information published by the Deutsche Nationalbibliothek
The Deutsche Nationalbibliothek lists this publication in the Deutsche Nationalbibliografie; detailed bibliographic data are available on the Internet at http://dnb.dnb.de.

ISBN 978-3-89673-722-9

Tel. +49 7045 930093 Fax +49 7045 930094
verlagwp@t-online.de www.verlagwp.de

Print and binding: Esser printSolutions GmbH, Bretten

Foreword

The Claus und Brigitte Meyer-Stiftung awards the Thomas-Gulden-Preis for outstanding academic achievements in memory of the former student Thomas Gulden who died from an incurable disease at the age of 25.

The Thomas-Gulden-Preis was awarded for the first time in the summer semester 2005. In the winter semester 2015/16, the Meyer-Stiftung awarded Tanja Schillinger for her outstanding academic results in the programme Business Administration/ International Business and particularly for her excellent bachelor dissertation with the Thomas-Gulden-Preis. In particular worthy of emphasis is her long-term voluntary engagement in different fields, for instance her work as voluntary tutor at the Familienzentrum of the Pforzheimer Stadtmission, and her cooperation in different university institutions during her studies at Pforzheim University.

The Claus und Brigitte Meyer-Stiftung publishes the dissertations of the laureates in the MEYER STIFTUNG monograph series. The scientifically comprehensive and sound bachelor dissertation of Tanja Schillinger addresses the highly-complex issue „An accounting approach to create an environmentally sustainable company: Selection and Definition of Environmental Indicators with special reference to Suppliers in Developing Countries“. This scientifically challenging topic has been elaborated with great care and through an analysis of copious literature. The increasing importance of the environmental friendliness of companies as a part of the economy should be highlighted. Hence, this dissertation is particularly relevant and explores new scientific territory.

The Claus und Brigitte Meyer-Stiftung is pleased to publish this valuable contribution to business theory and practice as Volume 18 of the monograph series.

We would like to thank sincerely Ms. Neugebauer and Dr. Brauner at the publisher Wissenschaft & Praxis for their generous support in publishing the monograph series.

Stuttgart, April 2016

Prof. Dr. Claus Meyer

MEYER
STIFTUNG

The Claus und Brigitte Meyer-Stiftung is a legally responsible and charitable foundation constituted under civil law based in Stuttgart, being approved as foundation by the administrative authority in Stuttgart on 21 April 2005. The purpose of the foundation is realized in promoting science and research, education and development as well as funding support of students in need at Pforzheim University, particularly by:

- Awarding the Thomas-Gulden-Preis for outstanding academic results and/or an excellent bachelor/master thesis in the area of Business, in particular Controlling, Finance and Accounting to one or several students.
- Awarding grants and similar support to students, particularly to those in need, to be able to continue and successfully accomplish their studies.

Thomas Gulden was born on 15 March 1978. He studied Business/Controlling, Finance and Accounting at Pforzheim University and completed his studies with a "very good" overall grade. Due to a congenital and developing myopathy, Thomas Gulden was confined to a wheelchair since the age of ten. On 11 April 2003, he died from this disease of which course he was aware. Posthumously, Thomas Gulden was awarded with an advancement award for his excellent diploma thesis, graded with a mark of 1.0. According to his wish, humanitarian organizations were supported with this award and his entire fortune.

Claus und Brigitte Meyer-Stiftung

Managing Director:

Professor Dr. Claus Meyer

Bernsteinstr. 102
70619 Stuttgart
Phone: 0711/4411488
E-Mail: claus.meyer@meyer-stiftung.de

Commercial Management:

Kathrin Zauner

Schulstr. 24
75385 Bad Teinach-Zavelstein
Phone: 07053/1432
E-Mail: sekretariat@meyer-stiftung.de

Bank details:

Sparkasse Pforzheim Calw

Account: 7670230
BLZ: 666 500 85
IBAN: DE 62 6665 0085 0007 6702 30

Website:

www.meyer-stiftung.de

Preface

This present paper was written as bachelor dissertation to gain admission to Bachelor of Science degree in the programme Business Administration/International Business at Pforzheim University. The thesis was submitted 30 July 2015 and comprises the corresponding knowledge on the subject at that time.

I would like to thank my supervisor Professor Dr. Joachim Paul for his support and advice. Thanks also go to Professor Dr. Harald Strotmann as my second marker. Their feedback and stimulating questions helped me to continuously find ways to improve my work.

A very special word of thanks goes to Prof. Dr. Claus Meyer, his wife Brigitte Meyer and the MEYER STIFTUNG for supporting me and publishing my academic work.

I dedicate this work to my family, in particular to my parents, Monika and Hans Schillinger, which have supported and accompanied me through my whole life. I would also like to thank my friends Michael, Daniel, Katrin, Stefanie and Anja. Their support as well as their useful feedback made my work what it is today.

Munich, April 2016

Tanja Schillinger

Table of Contents

List of Abbreviations

AHP	Analytical Hierarchy Process
BSC	Balanced Scorecard
BU	Business unit
CF	Carbon Footprint
CO_2	Carbon dioxide
DC	Developing country
DEA	Data Envelopment Analysis
ECI	Environmental Condition Indicator
ECP	Environmentally conscious purchasing
EF	Ecological Footprint
EHS	Environment, Health, Safety
EMA	Environmental Management Accounting
EMI	Environmental Management Indicator
EMS	Environmental Management System
EPA	Environmental Protection Agency
EPE	Environmental Performance Evaluation
EPI	Environmental Performance Indicator
ESCM	Environmental Supply Chain Management
EU	European Union
EuP	Energy-using Products
FA	Financial Accounting
FTE	Full-time equivalent
GDP	Gross domestic product
GHG	Greenhouse Gas
GJ	Gigajoule
GRI	Global Reporting Initiative
GSC	Green Supply Chain
GSCM	Green Supply Chain Management
HAP	Hazardous air pollutants
HRM	Human Resource Management
i.a.	Inter alia
ICSB	International Corporate Sustainability Barometer
IFAC	International Federation of Accountants
Incl.	Including
ISO	International Organization for Standardization
KE(P)I	Key Environmental (Performance) Indicator
KPI	Key Performance Indicator
LCA	Life Cycle Assessment

m^2	Square metre
m^3	Cubic metre
MA	Management Accounting
MEMA	Monetary Environmental Management Accounting
MF	Material Footprint
MPI	Management Performance Indicator
mt	Metric ton
mt of CO_2e	Metric tons of CO_2 equivalent
NGO	Non-governmental organization
ODS	Ozone-depleting substances
OECD	Organization for Economic Co-operation and Development
OPI	Operational Performance Indicator
PEMA	Physical Environmental Management Accounting
P&G	Procter & Gamble
PM	Particulate matter
POP	Persistent organic pollutants
QM	Quality Management
R&D	Research and development
RL	Reverse logistics
RoHS	Restriction of Hazardous Substances
ROI	Return on Investment
SC	Supply Chain
SCM	Supply Chain Management
SD	Sustainable development
SME	Small and medium-sized enterprise
TBL	Triple-Bottom-Line
tkm/ton-km	Ton-kilometre
UN	United Nations
VOC	Volatile organic compounds
WBCSD	World Business Council for Sustainable Development
WCED	World Commission on Environment and Development
WEEE	Waste of Electrical and Electronic Equipment
WF	Water Footprint
WRI	World Resources Institute

List of Figures

List of Indicators

This list shall give an overview of all selected indicators defined in chapter 6 by arranging them according to their respective value chain stage.

Inbound Logistics and Procurement (DC Suppliers) – Core Measures

Energy

Water

Material

Waste

Emissions

Costs

Optional Measures

Operations

Energy

Water

Material

Waste

Emissions

Corporate Environmental Indicators (KEIs)

1 Introduction

1.1 Problem Definition and Purpose

"As we watch the sun go down, evening after evening, through the smog across the poisoned waters of our native Earth, we must ask ourselves seriously whether we really wish some future universal historian on another planet to say about us: 'With all their genius and with all their skill, they ran out of foresight and air and food and water and ideas'" – U Thant[1]

Environmental issues have become ubiquitous, and so has the term sustainability.[2] A plethora of policies, companies, customers, investors, NGOs, and societies discuss sustainability; but talking and acting are two entirely different realms. The corporate sector, as one of the main polluters, bears a special responsibility to behave and act sustainably.[3] Many firms state that they care about sustainability.[4] However, sustainability is often not actually implemented into the company's culture and actions, but may represent a tool for looking good to the outside. In order to truly ingrain sustainability, there is a need to have specific internal instruments that track the firm's environmental performance.[5] This is where indicators come into the picture – more precisely environmental indicators. These are useful tools which may be applied by the accounting function in a firm in order to track its environmental performance and to support the decision-making of the management. However, in order to be useful they first have to be accepted and implemented, and unfortunately up until now, environmental accounting still seems to be rather an unpopular outsider in accounting.

To support the development towards the implementation of environmental aspects into accounting, this paper seeks to provide practical advice for German companies by selecting and defining a set of environmental indicators. The aim is to provide applicable and well-defined environmental indicators that address the most important global environmental issues, and that are assigned to the different value chain stages of a firm and beyond. This will be achieved, by also assigning indicators to the firm's suppliers, especially those suppliers from developing countries, as well as offering corporate environmental indicators.

---

1 The U Thant Institute (URL).

2 See chapter 4 and 2, incl. following sentence.

3 See Hawken (1996), p. 37; Herva (2011), p. 1687; Azapagic (2000), p. 243.

4 See section 2.2, incl. following sentence.

5 See chapter 5, incl. following sentences.

1.2 Delimitation and Research Approach

Great effort has been made to find measures, indicators and approaches towards sustainability.[6] It has started with the discussion, research and implementation of such means on a global, national and local sphere.[7] Yet, it has been recognized that it is impossible to go this journey towards sustainability without involving one of the main obstacles towards a sustainable world – the corporations.[8]

This present paper will therefore concentrate on the environmental aspect of sustainability from a corporate perspective; by providing generic environmental indicators[9], suitable for all kinds and sizes of companies in Germany. A special focus will lie on their developing country suppliers as they have become increasingly important to the supply chain and the paper seeks to encourage environmental sustainability in the entire chain. To be able to understand sustainability, and consequently the environmental indicators' true usefulness, application and relevance, the paper will first try to give a comprehensive overview of all three relevant thematic aspects: (environmental) sustainability, accounting and environmental accounting.

Therefore, the second chapter will present theoretical foundations concerning sustainability and related issues. As developing countries should be understood well before implementing corresponding indicators, their current awareness and practices of sustainability will be described in chapter 3. In order to understand the urgent need for change, chapter 4 will present current issues caused by unsustainable behaviour and actions with focus on businesses. Chapter 5 will finally present the last important theoretical basis for this paper by explaining important aspects of conventional and environmental accounting with special reference to indicators in the context of Germany.

In allusion to this, chapter 6 will clarify the significance of a holistic view concerning the creation of an environmentally sustainable company, and refer to Porter's value chain, which provides the basis for the selection and definition of the environmental indicators along this value chain with special reference to suppliers in developing countries. The selection approach is based on a comprehensive literature review, looking for the indicators most prevalent in theory and practice, that cover all important global issues, as well as on specific selection criteria. This chapter will

6 See Kates (2015), p. 802; chapter 5 and 6.

7 See Hahn (2006), p. 2; Hallay (1992), p. 17; Gallopín (1997), p. 13; Liverman (1988), p. 133; Atkinson (2000), p. 235; UNDP (2012), p. 7; for environmental indicators for the macro level see i.a. Wiedmann (2010); Bell (2008), pp. 29-40, 45ff; Wilson (2007); Hoekstra (2006); Wiedmann (2013).

8 See Hawken (1996), p. 37; Wolters (2002), p. 231; Azapagic (2000), p. 244.

9 Some indicators which deal for example with environmental costs, fines, investments and savings will also be provided. Some may perceive them as economic indicators, others as environmental indicators. In this paper they will be considered as environmental indicators (EMIs); see subsection 5.4.1; section 6.3.

conclude by offering not only specific divisional environmental indicators for each value chain stage but also presenting possible corporate environmental indicators most relevant for the management. Finally, it will give a brief introduction into the topic of an overall environmental indicator by demonstrating present issues associated with such measures, some opportunities and methodologies to derive a suitable indicator. Chapter 7 serves, on the one hand, as a summary of all selected indicators, and on the other hand, will deal with different models and frameworks to facilitate the implementation of the indicators. The 'Environmental Tree-Model' being proposed by the author. The last chapter closes with a critical evaluation and limitations concerning this work, as well as with opportunities and future prospects.

2 Sustainability

In order to understand the urgent need to address sustainability and to be able to follow with the right course of action, a firm has first to understand what sustainability means. Based on this, the current situation within industry concerning sustainability will be provided. Some related terms will be introduced: the shareholder and stakeholder approach, and the triple-bottom-line. Finally, the potential benefits will be highlighted, which can be achieved if a firm decides to tackle its environmental issues and starts its journey towards sustainability.

2.1 Definition and Differentiation

Sustainability and *sustainable development* (SD) are two terms which have been widely used and are continuously gaining recognition and popularity.[10] Many companies describe themselves as being sustainable, and support this declaration by generating sustainability reports.[11] Different stakeholders increasingly request firms to become sustainable.[12] It is not only in the corporate world that these terms are progressively being used; literature also focuses on sustainability and there is an increasingly vast amount of research in this field.[13] Surprisingly, sustainability science is globally spread, and it is not only industrialized countries[14] that are providing abundant, high quality input, but also developing countries (DCs).[15] Although sustainability and SD are commonly used, their meaning is still not entirely clear, and the concepts themselves are highly complex.[16]

No later than in 1992, SD gained international recognition during the 'Earth Summit' in Rio.[17] Although theories about the concept's meaning are inconsistent, there is consensus about the general idea of SD.[18] The report, *Our Common Future*, published by the WCED in 1987 provided this widely accepted definition of SD[19]: "[...]

[10] See Engelman (2013), p. 3f; Bennett (2013), p. 1f; Schaltegger (2000a), p. 46; Berns (2009), p. 3.

[11] See Öztürk (2014), p. 129; Zirkler (2011), p. 2.

[12] See IFAC (2005), pp. 7, 11, 25, 51; Schaltegger (2000a), p. 232f.

[13] See i.a. Kates (2015); Bettencourt (URL); Linnenluecke (2012).

[14] Industrialized and developed countries are used as synonyms in this paper.

[15] See Bettencourt (URL); for geographic distribution of sustainability science publications see Appendix 1.

[16] See Bennett (2013), p. 2; Hockerts (2001), p. 1; Rudawska (2013), p. 83; Azapagic (2000), p. 243; Ahi (2015b), p. 2884.

[17] See Hauff (2009), p. 1.

[18] See Kates (2015), p. 801; Aras (2009), p. 979; Liverman (1988), p. 133; Sikdar (2003), p. 1928; Hauff (2009), pp. 5–7; UNDP (2012), p. 7; Bennett (2013), p. 2.

[19] See WCED (1987), p. 41; Müller (2011), p. 25; also known as Brundtland Report see UNDP (2012), p. 7; Hauff (2009), pp. 5–7; Schröder (2014), p. 258.

development that meets the needs of the present without compromising the ability of future generations to meet their own needs."[20] SD may be seen as the process of reaching the long-term goal of sustainability.[21] However, a concrete distinction between sustainability and SD proves hard to define, and the two are therefore commonly used as synonyms.[22]

2.2 State of Corporate Sustainability

The definition of SD as laid out by the WCED is also applicable to businesses, resulting in the following definition for corporate sustainability: "[…] meeting the needs of a firm's direct and indirect stakeholders […], without compromising its ability to meet the needs of future stakeholders [...]."[23] Yet, in practice the meaning of the term has evolved and many firms define it differently.[24] The only agreement is that sustainability will have a large impact on corporations, but that this impact may differ from industry to industry.[25] The majority of companies state that they are already concerned with sustainability, however, this commitment seems often rather unstructured, and a clear strategy is lacking.[26] Although it has been under discussion for a lengthy period of time and has developed into a very important subject, the gap between the intention to do something and the effective implementation seems for many firms still insurmountable.[27] Many are setting goals and showing commitment, but struggle with the eventual execution, communication, and measurement.[28] Although there are numerous companies putting sustainability into practice, even the ones who are committed are still far from complete integration.[29] The leading companies are mostly made up of larger ones, as SMEs have still trouble putting sustainability into action due to financial restrictions, a lack of means, information and expertise.[30] Yet, they are making up some ground.[31] An encouraging development is that an increasing number of companies demonstrate an understanding of the bigger picture of sustainability by linking corporate actions to global issues. However, one

20 WCED (1987), p. 41.
21 See Clift (2000), p. 67; Ahi (2015b), p. 2882; WCED (1987), p. 41f.
22 See Aras (2009), p. 980; Ahi (2015b), p. 2882. Also used as synonyms in this paper.
23 See Dyllick (2002), p. 131, incl. whole sentence.
24 See Hockerts (2001), p. 1f; Linnenluecke (2012), p. 382f; Amini (2014), p. 13f.
25 See Berns (2009), p. 3; Robertson (2012), p. 1f.
26 See Berns (2009), p. 3; Baumgartner (2010), p. 76f.
27 See Eccles (URL); Berns (2009), p. 5; UNDP (2012), pp. 2, 4, 25f: On a national level; the increased relevance is also reflected by an increase in publications on this topic; see Linnenluecke (2012), pp. 382, 385; Appendix 2.
28 See UN Global Compact (2013), p. 12.
29 See UN Global Compact (2013), pp. 4f, 16f.
30 See UN Global Compact (2013), pp. 7, 16f; Berns (2009), p. 6; Doublet (2014), p. 1; Bergius (URL).
31 See UN Global Compact (2013), p. 7, incl. following sentences.

area which hinders a holistic corporate sustainable approach is a company's supply chain. Although many firms require their suppliers to align with their sustainable principles, very little has been done so far to track or support the suppliers' commitment.

With reference to Germany, the following can be stated according to a survey of the International Corporate Sustainability Barometer (ICSB).[32] It is usually assumed that German firms are strongly committed to sustainable issues.[33] Yet, when looking at the ICSB 2012, Germany's performance is rather average, despite its early awareness and favourable regulatory situation.[34] In addition to the Government, other driving forces include corporate networks and customers. Customer awareness is high but the acceptance for increased prices[35] of environmentally friendly products or tax increases is rather low. Recently though, the criticism of stakeholders concerning the consumption of resources, as well as emissions and waste has risen and puts pressure on German companies.[36] In general, Germany is slightly above average concerning the integration of sustainability into its core business. Environmental issues play a more important role than social issues, but with regard to the former, the accountancy function in particular is less affected by those issues.[37] The largest negative difference in "the engagement of functional units with sustainability measures"[38] between Germany and the international average can also be observed within the accountancy function. In Germany, the institutionalization of sustainability tools, like eco-indicators, is higher than in other countries, while environmental accounting lags well behind.[39] All in all, Germany is always around average; never particularly low on the scale, but it does not stand out either.[40] Reasons for this may be Germany's status as an export country, which depends on international developments. However, the most significant aspect is the implied ignorance of the accounting function concerning sustainability.[41] While a study in 2011 stated that sustainability has been ranked as one of the most important future topics of accounting by controllers, it lost its attraction somehow, and was featured some positions lower in a new survey conducted in 2014.[42]

---

[32] See Hörisch (2014).

[33] See Hörisch (2014), p. 93.

[34] See Hörisch (2014), p. 93f, incl. following sentences.

[35] Although customers in general may accept higher prices, see Öztürk (2014), p. 132.

[36] See Hörisch (2014), p. 98, incl. following sentence.

[37] See Hörisch (2014), p. 99f, incl. following sentence.

[38] Hörisch (2014), p. 100.

[39] See Hörisch (2014), pp. 100–102.

[40] See Hörisch (2014), p. 103, incl. following sentence.

[41] See Michel (2014), p. 98f; Horváth (2012), p. 45; Hörisch (2014), p. 100f.

[42] See Weber (2015), p. 17.

2.3 Shareholder Value vs Stakeholder Approach[43]

Corporate sustainability may also be linked to a shift in thinking from the shareholder value approach towards the stakeholder approach. According to the shareholder value approach the only thing that really counts is the satisfaction of the owners' needs.[44] That is, the maximization of the "shareholders' total return" [45] and the increase of the firm's stock price. With reference to shareholder value, the topic of agency theory may also be important.[46] Managers are actually often tempted to not act in the best interest of the shareholders, the principals, but in their own interest, the role of the agent.[47] The stakeholder approach goes one step further and aims to act in the best interest of all people with a stake in the firm.[48] Although the theory has evolved in different directions, the original stakeholder theory for strategic management was only meant to increase profits.[49] Freeman even refrains from describing social and ethical issues in regard to his idea of the stakeholder approach.[50] Despite not being originally devised to apply to sustainability, it can be an important thought-provoking direction; to not solely concentrate on the economic responsibility a firm has to its shareholders but an encouragement to focus on a wider picture.[51]

2.4 Triple-Bottom-Line

It is generally agreed that the Triple-Bottom-Line (TBL) comprises three dimensions, the social, the environmental and the economic dimension, and therefore "focusing on economic prosperity, environmental quality, and […] social justice."[52] The background is that firms should no longer only measure their economic performance, as they did with the shareholder and stakeholder approach, but also focus on their added environmental, ethical and social value.[53] Therefore, sustainability can also be

---

43 This topic will be touched briefly, for further reading see Rappaport (1986); Freeman (2010a).

44 See Rappaport (1986), pp. xiii, 1.

45 Rappaport (1986), p. 1, incl. whole sentence.

46 See Friedman (2007), p. 174f; Bennett (2013), pp. 14, 35.

47 See Rappaport (1986), p. 6.

48 See Freeman (2001), p. 10; Freeman (2010a), p. 25f.

49 See Freeman (2010b), pp. 10f, 29-31, 58f; Freeman (2010a).

50 See Freeman (2010b), p. 58.

51 See Freeman (2010b), pp. 40–42.

52 Elkington (1999), p. 70; see Norman (2004), p. 243; Elkington (2004), p. 3; Harrison (2013), p. 110; Elkington (1999), p. 70.

53 See Harrison (2013), p. 110; Elkington (2004), p. 3.

divided into three branches: environmental, social and economic sustainability.[54] The first focus of SD has primarily been set on the environment.[55]

2.5 Benefits

Sustainability is still seen by many as something that needs high investments that only large companies can afford, and it is often used as a mere marketing tool.[56] This is an untenable situation.[57] Although there may be a debate[58] about whether sustainability actually pays off, there are enough real life examples that sustainability will not hinder profits but scale them further – and can even become a strategy for competitive advantage.[59]

Sustainability may be assumed to be very expensive; needing high investments but only being profitable in the longer term.[60] Some may even argue that it can decrease shareholder value, but findings accumulate which refute this.[61] According to an analysis where high-sustainability companies were compared to low-sustainability companies over a period of 18 years, the results were clear.[62] The companies addressing sustainability did better with regard to their financial performance, including the stock market and other accounting measures; by gaining 4.8% higher return than the unsustainable companies. Missing the opportunity to jump on the bandwagon of sustainability may even lead to a competitive disadvantage in the future. Cost reductions especially are demonstrated through the reduction of waste and emissions, and new processes may lead to higher efficiencies.[63] Other more intangible benefits may be: an increased brand image and trust, as well as higher incentives for innovations and an increased response rate and flexibility to current trends.[64] The number of financiers aligning their investment on sustainability by making use of

54 See Danciu (2013), p. 8; Baumgartner (2010), pp. 78–80.

55 See Hauff (2009), p. 10; Elkington (1994); Elkington (1999), p. 70. As already mentioned this paper will also concentrate on the division of environmental sustainability that aims to protect the environment by reducing the negative impact of firms and improve the handling of natural resources; see Danciu (2013), p. 8.

56 See Hang (2015), p. 191; Bonini (2014), p. 5; UN Global Compact (2013), pp. 7, 17.

57 See Bonini (2014), p. 5.

58 For reading the opposite side of an early view see e.g. Walley (1994).

59 See Bonini (2014), p. 6; Berns (2009), p. 5; Hang (2015), p. 196; Rao (2005), p. 906. In the following there will be only a selection of possible benefits as a comprehensive analysis would exceed the scope of this work.

60 See Walley (1994), p. 46; Epstein (2008a), p. 26f; Stubbs (2008), pp. 515, 519; Dyllick (2002), p. 132; Berns (2009), p. 7; Bonini (2014), pp. 5, 9f; Haanaes (URL); van Bergen (2010), p. 7; EIU (2010), pp. 2, 6, 17.

61 See Eccles (URL); Sirkin (URL); Rudawska (2013), p. 84.

62 See Eccles (URL), incl. following sentences.

63 See Hang (2015), p. 192; IFAC (2005), p. 11; savings for SMEs see Bergius (URL).

64 See Rudawska (2013), p. 84f; Berns (2009), p. 5; UN Global Compact (2013), p. 4; IFAC (2005), p. 11.

sustainability indices[65] is growing too.[66] Not only investors but also other stakeholders such as customers and Governments will ask companies to act in a sustainable way.[67] Upcoming changes in laws and provisions will put unsustainable companies under pressure, while sustainable businesses are able to improve their relationship with key stakeholders. The shareholder value may also be increased through several benefits resulting from corporate sustainability, such as increased pricing power, cost savings, improvement of employee recruitment and productivity, increased market share, better ability to enter new markets and lower risks and costs of capital.[68] The savings may pay for future investments in environmental technologies.[69] In addition, the first mover advantages may be impossible to grasp for the firms who lag behind.[70] Being a pioneer or leader in the field of sustainability can certainly have some considerable benefits, yet nobody will argue that it is a great challenge.[71]

However, there are enough business cases which show that it is possible and pays off. For example, Bayer expected more than $10 million savings each year due to controls of their resource efficiency.[72] Lockheed Martin saved more than $7.5 million overhead by implementing some simple production improvements[73]. The assumption that a sustainability strategy only pays off in the long-term is clearly brought into question by firms that have demonstrated the opposite.[74] Interface, Inc., an American carpet producer, for example, reduced their total waste worldwide by 40% and made savings of $67 million in "the first three-and-a-half years of"[75] their sustainability programme. The same programme saved them $450 million between 1995 and 2013, which was only one initiative of many which further reduced costs and led to large savings.[76]

This small number of business cases demonstrates that sustainability can be a profitable strategy, and that overall it is "'the right thing to do'"[77]. Nevertheless, the stated examples refer to businesses of the developed world. The next chapter will examine the current situation in developing countries.

---

65 I.a. Dow Jones Sustainability Indices, Global Challenge Index, STOXX Sustainability Indices.
66 See UN Global Compact (2013), p. 4; IFAC (2005), p. 11; Berns (2009), p. 8.
67 See Berns (2009), p. 8; IFAC (2005), p. 11; Öztürk (2014), p. 132, incl. following sentence.
68 See Berns (2009), pp. 4–6; Appendix 3; Hang (2015), pp. 192, 196.
69 See Haanaes (URL); Anderson (1998), p. 11.
70 See Berns (2009), p. 8.
71 See Bonini (2014), p. 15; Haanaes (URL).
72 See Bonini (2014), p. 14, incl. following sentence.
73 Invested $240,000.
74 See Eccles (URL).
75 Anderson (1998), p. 11, incl. whole sentence; see Harel (2013), p. 33.
76 See Harel (2013), pp. 17f, 34, 38.
77 Epstein (2008a), p. 30.

3 Status quo in Developing Countries[78]

Suppliers from developing countries present an integral part of this work, as specific indicators will be assigned to them in chapter 6. For their effective use, it is essential to understand the concrete situation of suppliers in DCs. It shall become clear what a firm can expect and what they cannot expect from them, by showing the suppliers' awareness and practices in the field of sustainability.

3.1 (Environmental) Sustainability Awareness

Representatives of DCs often claim that a large number of their environmental issues are caused by developed countries.[79] This is backed up by some sources showing that a small portion of the industrialized countries cause the majority of pollution.[80] Despite this claim, it has also been suggested by empirical studies that DCs add greatly to pollution levels.[81] Moreover, DCs are slower to develop in the field of environmental sustainability in comparison to developed countries, since they put more emphasis on increasing their wealth and enhancing the well-being of their own society in order to catch up with industrialized states.[82] Some years ago, the assessment of DC companies as "agents of environmental sustainability"[83] was rather pessimistic.[84] It has been assumed that they are finding it particularly difficult to turn this situation around, as environmental regulations are not in place and policy often promotes business models that prevent SD.[85] The aforementioned drive to compete with developed countries and the pursuit of wealth has led to increasing environmental concerns.[86] However, the prevalence of these issues has emphasized the importance of addressing them; and in turn governments in DCs have begun to face the problem. As the President of the Republic of South Africa stated: "*'We have no choice but to develop a Green Economy.... There is significant opportunity for the development of a Green Economy in Southern Africa, and [...] to other parts of the continent*'"[87]. It is not only politicians who have recognized the urgent need for change; firms are also becoming

78 Developing countries refer to all countries which are less developed than industrialized countries concerning their economic, political and sociocultural situation; see Akamp (2013), p. 55; including the BRIC countries as well in this paper.

79 See Durth (2002), p. 103.

80 See Robins (1997): Results of the Earth Summit; SERI (2009), pp. 3, 17.

81 See Pulver (2007), pp. 191, 194.

82 See Haanaes (URL); Pulver (2007), pp. 191, 194; Michael (2014), p. 493f; Visser (2008), p. 489.

83 Pulver (2007), p. 191.

84 See Pulver (2007), pp. 191, 194.

85 See Haanaes (URL); Pulver (2007), pp. 191, 194; Visser (2008), p. 487.

86 See Michael (2014), p. 493f, incl. following sentence.

87 UNEP (2010), p. 06–07, incl. whole sentence.

more aware of the possible benefits and are progressing towards sustainability.[88] An increase in the awareness of and commitment to sustainability can be observed, as there are numerous examples of DCs who have begun successfully working towards sustainability, of which a small selection will be presented in the next section.[89]

3.2 (Environmental) Sustainable Practices

Industrialized countries have led the implementation of sustainability in business.[90] The approach of implementing sustainability practices into DC firms is often very different in comparison to those of industrialized countries.[91] DCs usually begin by taking small steps, while Western companies often take a more radical approach, with new and highly expensive technologies. Yet, sustainability practices among DCs can be observed "across Latin America, Africa, the Middle East, Asia and the South Pacific."[92]

Asia, for example, which was long known for its unsustainable practices, is in transition.[93] Due to more stringent listing requirements and the establishment of sustainable stock exchanges, Asian firms are moving in the right direction. China, in particular, is taking great steps within the field of renewable energy.[94] In 2005, a law was passed which provides Chinese companies financial incentives to invest in renewable energy. There is a considerable rate of action in the field of environmental reporting and supervision, but many firms are still lagging behind.[95] Two other examples are Shree Cement and Jain Irrigation Systems, both located in India.[96] Their investments in sustainability have resulted in lower energy consumption, high savings and competitive advantages. Concerning 'greening' the whole supply chain some Asian companies have made progress by offering their suppliers awareness seminars and transferring knowledge, and guidance on behalf of their own environmental programmes.[97]

Africa has also experienced a transformation towards more environmentally friendly practices. As early as 1977, the company Sekem pioneered in the field of organic

---

[88] See Haanaes (URL); UNEP (2010), p. 06–07; Pulver (2007), p. 194f.

[89] See i.a. UNEP (2010), p. 06–07; Pulver (2007), p. 194f.

[90] See Robins (1997).

[91] See Haanaes (URL), incl. following sentence.

[92] Haanaes (URL), incl. whole sentence. For further DC success stories see UNEP (2010).

[93] See Robertson (2012), p. 4, incl. following sentence.

[94] See UNEP (2010), p. 08–09, incl. following sentence.

[95] See Hang (2015), pp. 191f, 196.

[96] See Haanaes (URL), incl. following sentence.

[97] See Rao (2002), p. 650.

farming in Egypt.[98] It presented a new business strategy which paid off, not only environmentally but also financially. Uganda is also making great progress towards organic farming.[99] Kenya's Ministry of Energy created the 'Feed-in Tariff' which requires energy firms to buy their electricity from sources of renewable energy.[100] Consumption of renewable energy is also encouraged by the Tunisian Government in order to decrease their dependence on conventional energy sources.[101] In South Africa, Woolworths has implemented a balanced scorecard equipped with 'green' performance indicators.[102]

South American countries also contribute to this trend, through for example Brazil's launch of a Corporate Sustainability Index in 2005.[103] The company Natura in Brazil worked together with its suppliers to source only sustainably attained raw materials.[104] Masisa in Chile has greatly promoted its sustainable wood, and Florida Ice & Farm located in Costa Rica, saw huge savings by making minor repairs to leaks in their equipment. These companies all experienced higher profits and growth rates than their unsustainable competitors.

3.3 Important Considerations for the Control of DC Suppliers[105]

From the previous sections, it can be deducted that DCs in general are becoming more aware and more involved in the development of sustainable practices. Yet, in contrast to developed countries they are doing this little by little; since the means, availability of information and knowledge is lacking.[106] This should certainly receive some attention when asking for environmental data from suppliers in order to receive the respective indicators. Yet, as awareness is on the rise, it may become easier to implement environmental indicators for DC suppliers. However, there is still a huge gap between DCs and Germany, as the former often face political uncertainty and a lack of regulations in comparison to the high standards demonstrated in Germany.[107]

98 See Haanaes (URL), incl. following sentence.
99 See UNEP (2010), p. 12–13.
100 See UNEP (2010), p. 10–11.
101 See UNEP (2010), p. 22–23.
102 See Haanaes (URL).
103 See Lourenço (2013), p. 134.
104 See Haanaes (URL), incl. following sentences.
105 This chapter will not go into detail concerning supplier management. Further investigation on this topic can be found in subsection 6.4.2.
106 See Haanaes (URL); Akamp (2013), p. 56; Rao (2008), p. 36; Trapp (2014), p. 3; Gallopín (1997), p. 2; Weber (2014), p. 134f; Khan (2014), pp. 1212, 1220.
107 See Baskaran (2012), pp. 647, 656; accenture strategy (2015), p. 11: A survey comprising 3,396 companies worldwide; Visser (2008), p. 487; Robins (1997).

Furthermore, there are not only differences between DCs and developed countries to be aware of, but there are differences among DCs themselves, which should be addressed when implementing a control system on DC suppliers.[108]

Brazil, for example, has implemented sustainability regulations but its suppliers often lack awareness of environmental issues and therefore make only little effort concerning emission reduction goals, reporting and investment.[109] Yet, Brazil makes up a main part of the most important supply chains in the world inter alia: timber, livestock and coffee.

China, as demonstrated, is becoming increasingly aware of environmental issues. Yet, its suppliers receive the poorest scores concerning emissions reporting.[110] They perform below the global average in many fields, despite making recent progress.

Indian suppliers also face many problems, such as policy uncertainty and infrastructure issues, which have resulted in lower investments in sustainability.[111] Although India is facing many environmental risks, suppliers concerns are still relatively low and therefore their management processes lag behind.

When involving DC suppliers into one's own environmental sustainable practices, it may be important to consider the differences compared to own sustainable standards, possibilities, knowledge and means. The overall purpose should be to not only control the suppliers' sustainable practices but also support them in their progress towards sustainability, since some may not be able to do it alone.[112]

---

108 See accenture strategy (2015), in particular, pp. 4, 7 for an overview.

109 See accenture strategy (2015), p. 12, incl. following sentence.

110 See accenture strategy (2015), p. 16, incl. following sentence.

111 See accenture strategy (2015), p. 22, incl. following sentence.

112 See Robins (1997); Khan (2014), p. 1220; subsection 6.4.2.

4 Issues Caused by Unsustainable Behaviour and Unsustainable Business Practices

In order to persuade firms that a change for a better is inevitable, this chapter will explain the different issues caused by unsustainable practices, especially from a business viewpoint.

4.1 Underlying Thoughts

Humans and, in particular, firms often act in an unsustainable manner, leading to severe consequences.[113] Firms consume masses of natural resources and produce tons of emissions and waste along their entire value chain and beyond.[114] In the early stages of their procurement activities many buy materials, energy and water in order to process it in manufacturing. Of these materials, some become part of the product and some are only used for production purposes, resulting in waste and emissions. The products of firms often also cause severe environmental issues, after being disposed of into landfills. The problem lies not only with the end of the value chain, but as early as the first extraction of raw materials. Deforestation and mining of materials may affect the respective area negatively.

If these unsustainable behaviours continue in the same way in industrialized but also increasingly in DCs, in 1972 it was already predicted that the earth will reach its limit for growth within one hundred years.[115] This is no empty threat as the amount of resources extracted by 2030 may reach 100 billion tonnes.[116] The "resource-intensive economic model[,] Europe and other developed countries"[117] are pursuing will no longer be viable with regards to the environment as well as in a social or economic manner. Very important resources such as sand, water, air, forests, plants, oil, gas and rare earth, which cannot be restored or are difficult to restore, are fast depleting.[118] Society has developed to one which pushes the economy to harm the environment, since it measures success on the number of produced goods.[119] One should not forget that these products all carry an additional 'ecological backpack', meaning that the

[113] See Herva (2011), p. 1687; Tyteca (1996), p. 281; Hawken (1996), p. 37; Krajnc (2003), p. 279f; Formentini (2014), p. 1; Rao (2008), p. 24.

[114] See IFAC (2005), p. 20f, incl. following sentences.

[115] See Meadows (1972), p. 23; SERI (2009), p. 5.

[116] See SERI (2009), p. 3.

[117] SERI (2009), p. 5, incl. whole sentence.

[118] See Schmidt-Bleek (2014), p. 16.

[119] See Schmidt-Bleek (2014), p. 20.

price of a product does not reflect all the true costs.[120] The actual material input is much higher when taking into account all kinds of resources used. This demonstrates how humans make use of the planet in an unsustainable way; using it as both, a pantry and rubbish tip.[121] It is not possible to predict exactly what the outcome of such behaviour will be, but it is obvious that if the trend persists, there will be undeniable negative consequences.[122]

4.2 Environmental Issues

Many people may be aware of the impact their behaviour has on the environment, and yet it seems that there is a persistent attitude that the consequences will not be demonstrable to them personally. This is not always the case. In 2012 numerous extreme climate conditions were seen in Europe and the US, such as the immense amounts of snow at the beginning of the year, followed by severe droughts and the superstorm Sandy, which led to a plethora of personal injuries and damages to property.[123] Only recently, extreme weather also caused high damages in Germany, and such severe weather conditions are on the increase.[124]

These few examples demonstrate that environmental issues should be of common concern and taken seriously. In the same way, economic growth is progressively leading to further environmental damage.[125] Increasing global trade patterns and transportation, and the exorbitant use of fossil fuels have led to a large increase in greenhouse gas (GHG) emissions, which results in higher temperatures.[126] The deforestation which is carried out on a large scale has precarious impacts; such as changes in climate, which leads to less rainfall, resulting in an increase of deserted areas.[127] The ill-considered usage of nuclear energy has already led to areas being contaminated through waste and might have further impacts in future.[128] Despite the effort put into environmental protection, the world is facing increasing environmental issues such as: climate change (which is seen as one of the most urgent issues), a decline in biodiversity and resources, accumulation of aerosol in the atmosphere, acid rain, toxic pollution, severe weather extremes, waste (including increasingly e-waste),

---

120 See Schmidt-Bleek (2014), pp. 58f, 64, incl. following sentence.

121 See Schmidt-Bleek (2014), p. 113.

122 See Schmidt-Bleek (2014), p. 54.

123 See Writers (URL); Harvey (URL); Shore (URL); Guardian (URL); Russ (URL); for further current environmental issues see Singer (2013), pp. xv–xxi.

124 See Nieters (2015), p. 4f; FOCUS online (URL).

125 See Hunsberger (2012), p. 66.

126 See SERI (2009), p. 16; Durth (2002), p. 105f; Elliott (2006), p. 71f; Kuylenstierna (2012), p. 33.

127 See Schmidt-Bleek (2014), pp. 16, 31; Heinberg (2004), p. 297; Durth (2002), p. 108; Elliott (2006), p. 71; Hunsberger (2012), p. 88.

128 See Schmidt-Bleek (2014), p. 17; Elliott (2006), p. 71; Barra (2012), p. 170.

erosion, and the destruction of productive land, caused by carbon emissions and erosion.[129] Furthermore, the acidification of oceans, the scarcity of clean water, the pollution of water and soil, and the degradation of the ozone layer can no longer be overlooked.[130] The abundance of nitrogen can have further negative effects, such as coastal dead zones, meaning the "over-enrichment of waters with nutrients"[131], a reduction in the productivity of crop, forests and grassland, a loss of plants and animals, as well as climate change, leading to a loss of sea ice and Arctic warming, which may result in an increasing sea level and exposing large amounts of emission from the loss of permafrost.[132]

4.3 Drivers and Economic Issues

Environmental issues should not be neglected by corporations. However, it cannot be ignored that these issues can also have negative consequences with regard to a firm's economic situation. Sustainability drivers will also have an impact on firms, thus in the following, both drivers and consequences will be presented.[133]

The loss in global GDP each year is predicted to be at least one percent due to climate change.[134] The volatility and unpredictability of fossil fuel markets will also increase, as well as uncertainties concerning: the supply chain and the supply and production of goods.[135] The Government, certain 'watchdog groups' and the media are exerting more and more pressure on corporations.[136] In the case of the former, regulations concerning the environment are becoming stricter; take for example the reduction of pollution rights in the EU as of 2019[137] and eco-taxes[138]. This means that firms are forced to conduct more stringent controls and measurements of business processes, inputs and outputs, in terms of environmental protection. Some rather concerned countries have already established a wide range of environmental regulations, resulting in increased costs of compliance, control and clean-up of pollution, as well as emission fees.[139] It

129 See SERI (2009), pp. 3, 5, 10, 28; Schmidt-Bleek (2014), pp. 33, 38; Durth (2002), p. 106f; Elliott (2006), p. 71; Armenteras (2012), p. 134; Kuylenstierna (2012), pp. 33, 36; Barra (2012), pp. 168-170, 184; Gaddis (2012), p. 117f.

130 See SERI (2009), pp. 3, 5, 10, 17; Schmidt-Bleek (2014), p. 33; Elliott (2006), p. 71; Gaddis (2012), pp. 112, 114, 119; Kuylenstierna (2012), p. 33; Smith (2015c), p. 55.

131 Smith (2015b), p. 8; see Smith (2015b), p. 7f; Gaddis (2012), p. 109f.

132 See Alfthan (2015), p. 61; Gaddis (2012), pp. 117f, 119; Jäger (2012), p. 196.

133 See Boer (2012), p. 2.

134 See Boer (2012), p. 2; Kuylenstierna (2012), p. 36.

135 See Boer (2012), p. 1f.

136 See Bennett (2013), p. 2; Wiesehahn (2015), p. 29, incl. following sentences.

137 See Delhaes (2015), p. 8.

138 See Bartolomeo (2000), p. 42.

139 See IFAC (2005), p. 11.

can therefore be demonstrated that a driving force for a change in perspective is the shifting relation of costs for negative environmental impacts and costs for environmental accounting, while the first faces increasing costs the second becomes more profitable.[140]

Other stakeholders, such as communities, banks, insurance companies, customers and investors are also demanding green business practices and an increasingly number of firms have realized the potential monetary rewards in terms of savings, strategic benefits, as well as the growing market for green products and services.[141]

As the demand for resources increases due to the industrialization of DCs, some materials will become scarce and the number of trade restrictions will rise.[142] Population growth, water shortage and deforestation will put pressure on the agriculture industry, and water-intensive industries will compete for resources.[143] Water scarcity is another threat as it may lead to volatile prices and reputational issues, as well as create security risks due to possible conflicts.[144] In the past, ecosystem degradation was a focus topic due to businesses worrying about their reputation.[145] More and more companies recognize their dependence on a healthy ecosystem, as it directly influences access to and cost of resources. Deforestation, for example, is predicted to cost more than the financial crises in 2008.[146] In the future, firms will no longer be subsidised for certain inputs, and they will be forced through increasing regulations to pay for any damaging output, which is also known as the "polluter pays principle"[147].[148] These external environmental costs are likely to make up for a high percentage of firms' earnings, and for some industries it even could destroy their complete economic bottom line.[149] It is also conceivable that the whole financial system could become more unstable, resulting in inflation and deflation.[150] Yet, issues like resource scarcity may finally give rise to innovation opportunities, possibly resulting in a change to renewable resources and recycling.[151] If firms miss the chance to follow a SD, others will gain the competitive and cost edge, and those that are

140 See Schaltegger (2000a), p. 233.

141 See Öztürk (2014), p. 132; IFAC (2005), p. 11; Schaltegger (2000a), p. 232f; UN Global Compact (2013), p. 19.

142 See Boer (2012), p. 2.

143 See Boer (2012), p. 3.

144 See Boer (2012), p. 2f.

145 See Boer (2012), p. 3, incl. following sentence.

146 See Hunsberger (2012), p. 66.

147 Elliott (2006), p. 80.

148 See Boer (2012), p. 8; Rainey (2010), p. 245.

149 See Boer (2012), p. 9f.

150 See Heinberg (2004), p. 282f.

151 See Boer (2012), pp. 2f, 7.

unsustainable will risk high costs and loss of reputation.[152] This can even result in a competitive disadvantage, and may threaten a firm's existence.[153]

4.4 Other Issues

It is not only natural and monetary factors that one should be concerned about, but also the impact that unsustainable practices have on humans.[154] The negative impacts on health which are often influenced by output of firms are considerable.[155] Due to air pollution, millions of people die every year.[156] Additionally, many people experience lack of clean drinking water.[157] Water issues and other factors, such as environmental and infrastructural changes, often lead to diseases which have developed into major issues in some countries.[158] Unsustainable behaviours result also in social issues like "human rights violations, poor working conditions and low wages."[159] All this adds up to a decline in the quality of human life.[160] The shortage of resources may result in political, regional and industrial conflicts.[161]

These increasing issues should alarm laypersons, but also in particular the firms that are mainly causing these issues, as they can heavily contribute to their solution.[162] One possible contribution may be the implementation of sustainability in accounting departments, to push corporations in the right direction and to monitor their progress, which will be elaborated on in the following chapter.

[152] See Hopfenbeck (1993), p. 155f.

[153] See Stubenrauch (2014), p. 12; Hopfenbeck (1993), p. 156.

[154] See i.a. Kuylenstierna (2012), p. 36.

[155] See Elliott (2006), p. 48; Schmidt-Bleek (2014), p. 40; Rainey (2010), p. 243.

[156] See Kuylenstierna (2012), p. 33; Smith (2015a), p. 43.

[157] See Gaddis (2012), p. 114.

[158] See Gaddis (2012), p. 116f; Park (2015), p. 13f.

[159] SERI (2009), p. 10; see SERI (2009), p. 3; Elliott (2006), p. 52.

[160] See Durth (2002), p. 115.

[161] See Gaddis (2012), p. 125; SERI (2009), p. 28; Elliott (2006), pp. 46, 51f; Renner (2002), p. 152.

[162] See Hawken (1996), p. 37; Herva (2011), p. 1687; Azapagic (2000), p. 243; Robertson (2012), p. 1; Wolters (2002), p. 231.

5 Introduction to Environmental Accounting and Indicators

In order to work efficiently and effectively with the indicators stated in chapter 6, as well as to achieve the benefits from section 2.5, and to address the presented issues, this chapter will introduce the topic of environmental indicators which constitute a tool for environmental accounting. This will be achieved by providing a brief overview of conventional accounting and indicators. This is followed by the explanation of environmental (management) accounting which shall help to facilitate the understanding of the need and function of implementing environmental considerations into accounting. The ISO 14031 standard will be provided as a practical example of a guideline to measure a firm's environmental performance and thereby offer a number of environmental indicators. These indicators will be further explained in the last section, which will be concluded by an explanation of how the indicators have been selected for chapter 6.

5.1 Conventional (Management) Accounting

5.1.1 Classification and Tasks

Until today there is no clear definition in Germany of what accounting[163] actually comprises.[164] It is marked by change and development of the profession and tasks, and has experienced a steep increase in popularity.[165] Accounting can generally be divided into two broader categories, management accounting (MA) and financial accounting (FA).[166] Although both are somehow interconnected, FA concentrates on external reporting, providing financial information, while MA looks not only at monetary but also at non-monetary information, being the decision and control tool for internal management.[167] MA can be further divided into corporate accounting and business unit (functional) accounting.[168] There is also a difference between operative accounting,

---

163 In Germany, "Accounting" is different to "Controlling". In the English-speaking world the term "Accounting" is usually used, not "Controlling" being rather a German term. In Germany, "Accounting" and "Controlling" have different meanings. Yet, in this paper "accounting" will be used as a synonym to the German "Controlling".

164 See Hubert (2015), p. 1.

165 See Hubert (2015), p. 3f.

166 See IFAC (2005), p. 12; Schaltegger (2000a), p. 60f; Herzig (2012), pp. 11–14.

167 See IFAC (2005), p. 12f; Schaltegger (2000a), p. 60f; Herzig (2012), p. 11.

168 See Hubert (2015), p. 4f; Horváth (2009), p. 12.

focusing on 'hard facts' in the short-term and strategic accounting, which also looks at 'soft factors' for a long-term view.[169]

In Germany controllers have become real 'business partners' of the management.[170] Their role evolved from being just an information provider to a more strategic position.[171] A major part of their work still consists of the collection of needs-based information in order "to provide relevant, reliable and accurate information"[172] for management reports, and thereby supporting and guiding management decisions.[173] The accounting department is therefore a subsystem of management, i.a. enhancing transparency.[174] It can be concluded that accounting is not just control, as it might be assumed from the German term 'Controlling', but also reporting, tracking, recording, steering, guiding, regulating, commanding and coordinating, but also participating in the planning and decision-making process.[175]

For accounting, and derived from this, for management, it is important to measure things because only those can be steered.[176] If something has to be measured the question arises how to best track it.[177] For this, specific measures have been developed in accounting which will be explained in more detail in the following.

5.1.2 Indicators

5.1.2.1 Term and General Information

It is important to have strategic objectives and operational approaches in order to achieve and control the desired results, but this is not enough.[178] There is a need for continuous monitoring of corporate performance with the help of performance measurements. Starting with the collection and analysis of data in order to track progress, show off strengths and weaknesses, and eventually, to support enhancements. While a theoretical distinction between measures, metrics[179] and indicators is possible, they will be considered as synonyms as it is often done in

---

169 See Müller (2011), p. 21; Probst (2006), p. 22.

170 See Berlin (2014), p. 47; Colsman (2013), p. 47; Horváth (2012), p. 44; or 'sparring partners': Horváth (2009), p. 17.

171 See IFAC (2005), p. 13.

172 Schaltegger (2000a), p. 76, incl. whole sentence.

173 See Gladen (2014), pp. 3, 8; Hubert (2015), p. 2.

174 See Hubert (2015), p. 29.

175 See Horváth (2009), pp. VI, 15-17, 61f, 64; Herzig (2012), pp. 7–10.

176 See Hubert (2015), p. 29.

177 See Schaltegger (2000a), p. 60f.

178 See Franceschini (2007), p. vii, incl. following sentences.

179 The term metrics will also be used for measurement units.

practice.[180] In general, they are a "measurable representation of the condition or status of operations, management, or conditions"[181], presenting an essential instrument of accounting[182].

Indicators are very versatile and they pursue different purposes.[183] They can be used for internal or external purposes,[184] representing operational facts and the firm's (health) state.[185] Indicators are considered as fundamental to all corporate functions, they educate the employees by showing what is important and they give directions by presenting possible gaps between objectives and the current state.[186] They may have a motivational impact on employees as their performance gets tracked and the objectives are clear.[187] In addition, indicators possess a control function, they facilitate communication, showing opportunities for improvement and serving as excitation.[188] They can also reduce possible negative effects of the agency theory as the principal gains control over the agent.[189]

There are some important considerations when making use of indicators.[190] They should provide fast and accurate data, and they should be based on a firm's strategy and strategic objectives.[191] Their objective should be clear, and they should ideally encompass past, present and future information.[192] All in all, they should have an informative nature, and give a fast but comprehensive overview in a simple manner.[193] Firms should make active use of them by truly implementing them in daily work, and assign a responsible person to each indicator.[194] The indicators should be provided to the right contact persons.

There is often a system of indicators with several levels of aggregation.[195] Since managers usually lack time, they should only receive high-level indicators. Only in

180 See Franceschini (2007), pp. 12, 60; Müller (2011), p. 59; Gladen (2014), p. 9f.

181 DIN (2013), p. 9.

182 See Probst (2006), p. 9; Colsman (2013), p. 63.

183 See Probst (2006), p. 14.

184 This paper focuses on the internal accounting, and therefore on the internal use of indicators.

185 See Probst (2006), p. 14; Franceschini (2007), p. 1; Reichmann (2011), p. 23f; Gladen (2014), p. 10.

186 See Franceschini (2007), p. 9; Probst (2006), p. 14; Wolters (2002), p. 244.

187 See Gladen (2014), p. 29; yet, this should be seen critical: Czymmek (2001), p. 10.

188 See Franceschini (2007), pp. 10f, 14; Gladen (2014), p. 26.

189 See Franceschini (2007), p. 19; Gladen (2014), p. 23f; Bennett (2013), p. 14.

190 In subsection 5.4.3 complementary criteria will be provided which indicators should fulfill and which are important for the selection process.

191 See Hopfenbeck (1993), p. 331; Franceschini (2007), pp. 8f, 14, 101; Wolters (2002), p. 233.

192 See Probst (2006), p. 18.

193 See Reichmann (2011), p. 24; Hopfenbeck (1993), p. 331.

194 See Probst (2006), p. 18f, incl. following sentence.

195 See Gladen (2014), pp. 11–13, incl. following sentences.

emergencies, more detailed low-level indicators will be provided (management by exception). So there should be specific indicators for each hierarchy level.[196] Firms usually have an indicator data sheet which comprises some generic indicators that are widely used in all kinds and sizes of firms although there are and should be individual adaptions to own needs.[197] The frequency of indicator changes should be rather low in order to become a routine.

Concerning the approach to the final implementation and use of indicators, the first considerations concern the formulation of objectives and strategies.[198] Next is the selection of appropriate indicators, their frequency of creation (monthly, quarterly, etc.), and where to obtain data from, e.g. from bookkeeping, technical and quality management.[199] After coming up with a first design of an indicator data sheet, which will be discussed and checked for issues, it will be presented to the directors and decided over responsibilities of creation, distribution and analysis.

Selecting the right indicators is not a simple task because a firm becomes and is what it measures but sometimes does not get what it intended.[200] As an example, if a firm measures indicators 'a, b, c' and not 'x, y, z', it neglects the latter, and only the former may be reached at the expense of the latter, which may lead to counterproductive behaviour.[201] Thus, the most difficult part of implementing the right indicators is to identify those that constitute the process the best: "the so-called *Key Performance Indicators (KPI)*."[202] The most critical aspects in the process are: the definition, the acceptance and understanding on behalf of managers and employees, and the traceability and verifiability.[203] There are often a large number of indicators but only a few are used and the responsibilities are often confusing.[204] This should be avoided. Twenty high-level indicators may be sufficient but it may be feasible to possess hundreds or thousands of metrics.[205]

---

196 See Probst (2006), p. 14f.
197 See Probst (2006), p. 20f, incl. following sentence.
198 See Gladen (2014), p. 25.
199 See Probst (2006), p. 21, incl. following sentence.
200 See Franceschini (2007), pp. 2, 73; Brown (1996), p. 28; Probst (2006), p. 49f.
201 See Franceschini (2007), p. 2; Gladen (2014), p. 23; Probst (2006), p. 49f.
202 Franceschini (2007), p. 7, incl. whole sentence.
203 See Franceschini (2007), p. 8.
204 See Probst (2006), pp. 16f, 49.
205 See Brown (1996), p. 4.

5.1.2.2 Categories of Indicators[206]

There is the general distinction between absolute and relative indicators.[207] Absolute indicators comprise single values, sums, differences and mean values. Relative indicators present ratios, further divided into structural (same population in numerator and denominator), relation (different population in numerator and denominator) or index numbers (relative change of data over time).[208] Another classification is the distinction between leading and lagging indicators.[209] Leading indicators support early detection, while lagging indicators often offer information too late. Indicators also differ in their quantifiability, resulting in hard and soft indicators.[210] Hard indicators are easily quantifiable, while soft indicators focus more on qualitative factors. The goal is to quantify soft indicators to make them measurable. When classifying indicators according to their means, there are two important types: content and performance indicators.[211] When referring indicators to the TBL, they result in environmental, economic and social indicators.[212]

5.1.3 Indicator Systems

"Indicator systems consolidate single indicators into a system of dependent and complementary metrics."[213] As it is already the task of single indicators, a system even enhances the delivery for decision-oriented information through the adapted level of aggregation of single values, and the overall goals becoming more precisely to lower levels.[214] It allows for better alignment and coordination.[215] Alignment aims to provide "consistency between the strategic goals and indicators" for lower levels, while coordination supports the consistency between the indicators. An indicator system also serves as an early warning system through internal analyses, having defined tolerance areas.[216] Real issues can be recognized through comparisons between different periods, plan and actual, and between companies.[217] There are different architectures for indicator systems, such as 'arithmetic systems' which are built up of breakdowns in

206 There will only be a small selection of the categories considered as most important for the paper.

207 See Goldmann (2001), p. 597f; Gladen (2014), p. 14, incl. following sentence.

208 See Goldmann (2001), p. 597f; Probst (2006), p. 13; Gladen (2014), pp. 15–17.

209 See Gladen (2014), pp. 36, 53, incl. following sentence.

210 See Probst (2006), p. 22f, incl. following sentences.

211 See Sikdar (2003), p. 1930. Content indicators: In 5.3 referred to as condition indicators.

212 See Sikdar (2003), p. 1930.

213 Müller (2011), p. 59: "Kennzahlensysteme fassen Einzelkennzahlen zu einem System gegenseitig abhängiger und sich ergänzender Größen zusammen."

214 See Hopfenbeck (1993), p. 330.

215 See Franceschini (2007), p. 13f, incl. following sentence and quotation.

216 See Reichmann (2011), pp. 31–33; Probst (2006), pp. 14, 36; Herva (2011), p. 1687; Dale (2001), p. 3f.

217 See Probst (2006), p. 36; Gladen (2014), p. 103f.

numerators and denominators, substitutions and extensions.[218] Yet, such systems have many auxiliary key figures with little significance and do not show empirical impacts.[219] A 'classification system' can overcome those limitations by making use of non-mathematical indicators and an order by topics which offers higher flexibility, but allows for a higher degree of subjectivity.[220] These different approaches have led to different systems used in practice, such as the DuPont system, Tableau de Bord, ZVEI-System, RL-System, Critical Few Method, or the Balanced Scorecard.[221]

5.2 Environmental (Management) Accounting

As the acceptance of the need for change concerning the environmental dimension of the TBL has grown in recent years, criticism of conventional accounting has appeared.[222] Although the environment is suffering due to unsustainable business, the required information is often not provided by the accounting function. Environmental resources and impacts often do not show up in balance sheets and accounts, external costs are not internalized, and ecological issues only become obvious if firms are faced with penalties, licenses and clean-up costs. Economic-financial performance indicators from conventional accounting put emphasis on costs, lack considerations in quality, innovation and strategy, prevent the required development and are mostly lagging indicators.[223]

Although a high number of firms are already engaging in external environmental reporting following, for example, the standards of the Global Reporting Initiative (GRI) or the IR-Framework, sustainability is lagging behind in terms of internal reporting.[224] Therefore, firms which are already very active in external reporting, but not in internal management, may be at a disadvantage since they report about their environmental situation but have no instruments to enhance it.

Environmental management is still primarily done by separate environmental divisions on an ad hoc basis.[225] It is often assumed that sustainability and accounting are not compatible.[226] Yet, controllers should not have any reservations about new

---

[218] See Gladen (2014), p. 98.

[219] See Gladen (2014), p. 99f.

[220] See Gladen (2014), p. 100f.

[221] See Probst (2006), pp. 33–36; Franceschini (2007), p. 13; Gladen (2014), p. 98f; also for further information on this topic.

[222] See Bennett (2013), pp. 3, 5; Schaltegger (2000a), pp. 76–78, incl. following sentences.

[223] See Franceschini (2007), p. 16.

[224] See Stubenrauch (2014), p. 13, incl. following sentence.

[225] See Berlin (2014), p. 47; Bennett (2013), pp. 3, 9, 18; Colsman (2013), p. 45.

[226] See Colsman (2013), p. 44.

instruments.[227] Especially as there are already a number of present instruments such as eco-auditing or life cycle assessment (LCA)[228], and it is not only new information that is needed but also existing information[229]. As business partners, controllers should support the development and operationalization of environmental sustainable strategies.[230] They should certainly play an integral part of environmental management due to their superior knowledge of collecting, analysing, measuring and communicating information.[231]

Environmental accounting can be divided in terms of its content: into financial or non-financial data; and its purpose: external or internal.[232] This leads to four different approaches: financial reporting, social accountability reporting, energy and materials accounting and environmental management accounting (EMA).

This section deals with EMA. Although there is no official definition, IFAC states that it is "'the management of environmental and economic performance through the development and implementation of appropriate environment-related accounting systems and practices. While this may include reporting and auditing in some companies, environmental management accounting typically involves life-cycle costing, full-cost accounting, benefits assessment, and strategic planning for environmental management.'"[233] In Germany this accounting form was developed with a focus on eco-balancing.[234] It may represent an expansion of the traditional accounting or a completely new approach but it serves for internal decision-making.[235] EMA encompasses Monetary Environmental Management Accounting (MEMA) and Physical Environmental Management Accounting (PEMA).[236] MEMA is concerned with the environmental impacts on an enterprise which can be described in monetary units. PEMA expresses the environmental impact a firm has in terms of physical units. For the latter, it is essential to collect correct physical information concerning all inputs and outputs.[237] All material flows should be tracked through all processes in which materials are involved, which is also known as "'materials flow accounting'"[238].

227 See Müller (2011), p. 32, incl. following sentence.

228 See section 6.1.

229 See Bennett (2013), p. 3.

230 See Berlin (2014), p. 47.

231 See IFAC (2005), p. 11f; Bennett (2013), p. 30f.

232 See Bartolomeo (2000), p. 32f; IFAC (2005), p. 13; Appendix 4, incl. following sentence.

233 IFAC (2005), p. 19, incl. whole sentence.

234 See Bartolomeo (2000), p. 47.

235 See Bennett (2013), p. 3f; Schaltegger (2000b), p. 12.

236 See Burritt (2002a), pp. 26f, 29f; Schaltegger (2000b), pp. 14–16; Burritt (2002b), pp. 5f, 8f, incl. following sentences. See Appendix 5.

237 See IFAC (2005), pp. 21, 30.

238 IFAC (2005), p. 31, incl. whole sentence.

This can be extended to the whole supply chain by including physical data from suppliers and customers.[239] It is important to include both physical and monetary indicators, as the production managers are rather concerned with physical tools while executives prefer monetary and aggregated environmental performance indicators (EPIs).[240]

There are several challenges to EMA. Due to limitations in the conventional management accounting systems it may become difficult to collect accurate environmental information.[241] This may lead to false data, and therefore to misinterpretations and bad decisions. The environmental department usually possesses good knowledge about environmental issues, while technical staff could provide important information about flows of energy, water and others. Yet, both parties often do not know how this will be translated into accounting information. Controllers are experts in accounting but do not know much about environmental issues and flows, thereby they cannot provide the needed accounting information. The provided environmental data should be essential to the firm's specific situation, but is often not accurate or adequate, or incomplete, due to missing overhead accounts or future environmental costs.[242] One should also be aware of trade-offs, i.e. if a measure is realized to improve environmental results it may lead to other negative environmental consequences.[243] However, if an accounting system is implemented it may help to improve the situation in different manners, improving the formal correctness, the completeness, the consistency, and the content validity of environmental issues.[244]

5.3 Environmental Performance Evaluation – ISO 14031[245]

Even though there is a continuous discussion about environmental indicators, only a small number of considerable guidelines exist, ISO 14031 being the most significant.[246] Environmental Performance Evaluation (EPE) was set forth by the International Standard in its ISO 14031 that provides guidance for all kinds of firms to assist them in the measurement, evaluation and communication of their environmental performance and to achieve environmental performance objectives with the help of

239 See IFAC (2005), p. 32.

240 See IFAC (2005), p. 41; Burritt (2002a), p. 33f; Gladen (2014), p. 12. These management indicators in monetary units will hereinafter be referred to as EMIs; see subsection 5.4.1.

241 See IFAC (2005), p. 26, incl. following sentences.

242 See Bennett (2013), p. 26; IFAC (2005), p. 27f.

243 See Bennett (2013), p. 29.

244 See Bennett (2013), p. 31.

245 This section will only give an overview of ISO 14031. For further details see DIN (2013).

246 See Scherpereel (2001), p. 100; Danse (2003), p. 43; Haubach (2013), p. 32.

KPIs, or key environmental performance indicators (KEPIs).[247] It is structured according to a "'Plan-Do-Check-Act' [..] management model"[248].

EPE makes use of two different types of indicators, environmental condition indicators (ECIs) and EPIs.[249] ECIs present the overall situation of the environment likely to be affected by an organization, such as the regional air quality[250]. Yet, it is hard to assign ECIs to specific corporate operations and they are primarily used by public institutions.[251] But they can help firms with the selection of EPIs, as they reveal the company's most important environmental issues.[252] EPIs present information with regards to the management of the firm's relevant environmental issues as well as showing the results of environmental initiatives.[253] EPIs are further divided into management performance indicators (MPIs) demonstrating management efforts in order to enhance environmental performance, and operational performance indicators (OPIs) which show the environmental performance of operational processes.[254] The standard proposes to "select a sufficient number of relevant and understandable indicators"[255]. ISO 14031 also gives information on and examples for the selection of the different types of indicators.[256] Many more examples of indicators are provided in the annex.[257]

Although ISO 14031 is conceivably the best tool to develop an environmental indicator system, there is also criticism surrounding it.[258] One critical aspect is its lack of any prioritization for the indicators' development, leading to a high degree of complexity.[259] Due to its complexity it may not be appropriate for true beginners or smaller firms.[260] It also neglects standardization, support for implementation, external communication, and an overall connection to SD. That said, ISO 14031 is an international standard that tries to find consensus over a large scope.

---

247 See DIN (2013), pp. 4-6, 10f; Scherpereel (2001), p. 100; Pérez-Calderón (2012), p. 1008; Singhal (2004).

248 DIN (2013), p. 11, incl. whole sentence. For details see DIN (2013), p. 11f.

249 See DIN (2013), p. 12, incl. following sentence.

250 See Jasch (2000), p. 83.

251 See DIN (2013), p. 12; Jasch (2000), p. 83.

252 See Jasch (2000), p. 83.

253 See DIN (2013), p. 12.

254 See DIN (2013), pp. 9, 12.

255 DIN (2013), p. 18, incl. whole sentence.

256 See DIN (2013), pp. 22–33.

257 See DIN (2013), pp. 45–70.

258 See Danse (2003), p. 43; Scherpereel (2001), p. 100.

259 See Scherpereel (2001), p. 100; Bennett (1999).

260 See Bennett (1999), incl. following sentences.

5.4 Towards Environmental Indicators

5.4.1 General Information

Environmental indicators are assumed to be the most appropriate instrument to consider environmental aspects in accounting.[261] Environmental indicators are quantifiable measures to assess, record and communicate environmental conditions in terms of physical measures.[262] The implementation of environmental indicators comprises different objectives, i.a. reducing input, resource use, emissions, waste and traffic, while increasing the share of renewable energy and recycling, as well as trying to develop environmentally friendly product alternatives and securing economic success.[263] With the help of indicators changes in environmental performances can be detected.[264] As they offer a basis for improving the environmental situation, they may also result in financial and competitive advantages.[265] The firm can gain a better awareness of environmental regulations. It supports the development of corporate environmental objectives and employees may become more motivated to follow a strategy of reducing resource consumption. Other benefits are the increase in transparency and general awareness.[266] Eventually, they also support long-term thinking.[267]

In order to track the indicators' progress in a meaningful manner and to drive performance, it is essential to set environmental objectives and goals, and/or orientate on standard values and legal limit values.[268] There are three different kinds of environmental indicators, here referred to as ECIs, EPIs and environmental management indicators (EMIs)[269].[270] This paper will focus on the second category of indicators which help to plan, steer and control the environmental impacts a firm has; and on EMIs which do not provide information regarding the environmental performance or impacts but they do help to quantify "environmental management targets"[271], and show management efforts promoting environmental sustainability[272].

---

261 See Zirkler (2011), p. 2.

262 See Schaltegger (2000a), p. 299.

263 See Goldmann (2001), p. 598f.

264 See Rao (2014), p. 63.

265 See Schaltegger (2000a), p. 303, incl. following sentences.

266 See Hansgrohe SE (E-Mail).

267 See Schaltegger (2000a), p. 304.

268 See OECD (2011), p. 43; Günther (2002), p. 223; Wolters (2002), p. 244; Brown (1996), p. 9; Müller (2011), p. 61.

269 In ISO 14031 also referred to as a kind of EPIs: MPIs; see section 5.2. EMIs and MPIs will be considered as comprising the same in this paper.

270 See Goldmann (2001), p. 600f, incl. following sentence.

271 Jasch (2000), p. 83, incl. whole sentence.

272 See DIN (2013), p. 12. According to DIN (2013), p.22, MPIs comprise "managing matters such as training, legal requirements, resource allocation [...], environmental cost management, purchasing, product development [...]".

The indicators are usually dependent on different factors, e.g. the industry, management approach, corporate culture, regulatory framework, technology, size and structure, and therefore differ among firms.[273] Although specific indicators are important, generic indicators will be defined which provide information about common environmental media that all firms should track.

Another group of indicators considers not only the environmental dimension of the TBL but also the economic, known as eco-efficiency indicators, which combine economic with environmental performance.[274] Thus, economic figures (numerator) measured in monetary terms and ecological figures (denominator) measured in physical terms will be combined to efficiency indicators.

5.4.2 Possible Challenges

Although smaller firms commit themselves to sustainability, they may dismiss concrete actions concerned with environmental improvement as costly and realize the required knowledge is missing.[275] The collection of all the necessary data for environmental indicators may also seem to be problematic but some important data is already sufficiently available through existent internal records,[276] making it possible to lower costs by just adapting the current system to the new requirements.[277] However, it is important here that the environmental information procurement does not exceed the potential benefits.[278] The adaption of the indicators to the different levels of aggregation may also represent a challenge but is important in order to be relevant for all addressees.

In practice, a number of "common and sectoral indicators" [279] are already provided for firms which want to start to conquer their environmental issues. However, many indicators evolved from limited research and may not be easily applicable. Firms will face additional practical problems when engaging in the development of EPIs and EMIs, such as making use of only a few and significant indicators, and those that are easy to understand. There should be sufficient flexibility as no generic indicator system will apply to all companies perfectly. In addition, firms should pay attention to quantifiable physical performance indicators and remain able to connect them to conventional performance indicators. The most important elements of the value chain have to be identified and a LCA might be necessary.

---

[273] See Schaltegger (2000a), p. 299; Jasch (2002), p. 48f, incl. following sentence.

[274] See Schaltegger (2000a), p. 361f; Verfaillie (2000), p. 3, incl. following sentence.

[275] See UN Global Compact (2013), p. 16f.

[276] See Hopfenbeck (1993), p. 337; Zirkler (2011), p. 3; Goldmann (2001), p. 597.

[277] See Schaltegger (2000a), p. 56f; Wolters (2002), p. 244f.

[278] See Schaltegger (2000a), p. 56f, incl. following sentence.

[279] Schaltegger (2000a), p. 304; see Schaltegger (2000a), p. 304f, incl. following sentences.

5.4.3 Selection and Selection Criteria

When the decision has been taken to establish a system of environmental indicators the first questions that arise may be which indicators should be chosen and what are the criteria laid down for this non-marginally choice. The following criteria, if possible, will also be applied for the choice of indicators in chapter 6.

The selected indicators should be measurable, simple (especially for the start), easily assessed, relevant/important (to all addressees, key objectives and the corporate environmental strategy), useful (for supporting the achievement of the environmental performance objectives, readily understood and provide guidance), economic (provide more benefits than costs, information collection not too costly and time-consuming, reasonable administrative effort), representative (concerning the environmental performance), responsive (to changes), robust (for time and firm comparisons, understood similarly by its users), should consider availability of required data, should have the right degree of detail and offer "behavioural soundness"[280] (provokes only the wanted actions, no counter-productive ones).[281] The indicators do not necessarily meet all the mentioned criteria to be considerable for the firm, as there are also some trade-offs between certain criteria.[282]

Before the implementation of an indicator system, several questions should also be answered by the firm.[283] The company should think of whether…

- …sufficient data is already available.
- …new data is necessary.
- …the data available is also sound today and in future to ascertain future comparability.
- …the indicators will support managers' decision-making…
- and what will be the right level of aggregation and how the indicators become comparable.

It is also of importance for the selection process to adapt the indicators to the available information, which can be obtained through different assessment tools such as the ecological balance sheet or LCA.[284] They offer first insights into absolute values concerning different environmental issues, such as consumption of materials, energy, water, and the generation of waste and emissions, which are most significant

---

280 Caplice (1994), p. 14.

281 See DIN (2013), p. 49f; Caplice (1994), p. 14; Erol (2011), p. 1089; Wolters (2002), p. 244; Handfield (2002), p. 77; Probst (2006), p. 52f.

282 See DIN (2013), p. 50; Caplice (1994), p. 16.

283 See Schaltegger (2000a), p. 304, incl. following key points.

284 See Hopfenbeck (1993), p. 330; Benedetto (2009), p. 901; Olsthoorn (2001), p. 456; LCA see section 6.1.

concerning the environmental impact.[285] To begin with, the first chosen indicators should address the main environmental issues and impacts a firm is facing.[286] For specific indicators the firm can define a scale with a minimum and maximum value for each indicator to put it into perspective.[287] All in all, "[t]he most promising approach in the early stages of implementation is to develop simple, readily understood measures that reflect environmental priorities and make maximum use of available data."[288]

Concerning the selection approach for the indicators of chapter 6, they have been chosen on the following basis. The author has conducted an extensive literature review, where a high number of mainly environmental indicators have been collected. These indicators have been listed by their frequency in theory but also according to their applicability stated in the literature, which has been determined through different approaches[289]. Moreover, a main focus has been set on the environmental indicators stated by the GRI, as this standard is the most-commonly used worldwide (e.g. 82% of the world's 250 largest companies[290]).[291] This was also assumed to be an indication for the applicability of its proposed indicators. An attempt was made to only choose the most general indicators and forego the more specific ones, since most references, such as the GRI, have proposed too many indicators. For selection it has also been tried to include all global environmental issues into the indicators. In order to limit the number of possible indicators, combined indicators which include two or three dimensions of the TBL have been excluded, such as eco-efficiency indicators. When referring to the stated selection criteria, the aim is to find the most: relevant, general, applicable, quantifiable and easy to use and understand environmental indicators[292] that refer best to one of the value chain stages, addressing the most important environmental issues.

---

285 See Hopfenbeck (1993), p. 330; IFAC (2005), p. 32.

286 See Wolters (2002), p. 233.

287 See Müller (2011), p. 63.

288 Wolters (2002), p. 244.

289 Such as Delphi studies, expert interviews or approaches mentioned in subsection 6.4.2.2.

290 KPMG International (2013), pp. 11, 30f: Also widely used in Germany.

291 See Toppinen (2013), p. 203; Brown (2009), p. 1f; Levy (2010), p. 88f; KPMG International (2013), pp. 11, 30f; GRI is intended for external reporting, but in order to be able to report those data, they should be developed but in particular controlled and steered internally. Moreover, the GRI indicators also encompass all important global environmental issues which are essential for the internal reporting as well. Therefore, GRI has been assumed to be an appropriate source.

292 Comprising both: EPIs and EMIs.

6 Definition of the Selected Environmental Indicators along the Whole Value Chain

The theoretical foundations for this chapter have been laid with the previous chapters that explain sustainability, accounting and consequently environmental accounting and indicators. This basis has been supplemented with the information about DCs which will be important for the subsection 6.4.2. Before that, two aspects; environmental supply chain management and LCA will be briefly explained, as they can support the work of environmental accounting. The order in which the indicators will be presented will be based on Porter's value chain units, which will therefore be presented as well. The structure of this chapter will be elaborated in section 6.3. Section 6.4 defines the indicators for the different value chain stages, starting with the indicators for suppliers in DCs, followed by those for procurement and inbound logistics. The indicators for product design will be next as those are located at the beginning of the value chain. The two following functions, firm infrastructure and HRM relate to the whole firm. The indicators for operations, outbound logistics, marketing, sales and service close the section for specific divisional environmental indicators. These specific indicators will be summarized to key environmental indicators and the chapter concludes with a theoretical background of a possible single environmental indicator.

6.1 Environmental Supply Chain Management and Life Cycle Assessment

When defining and selecting indicators for a firm to make it truly environmentally sustainable, the focus should not only be put on the internal value chain but go beyond this. In this context, the terms Supply Chain (SC), Green Supply Chain (GSC), Supply Chain Management (SCM), Environmental or Green Supply Chain Management (ESCM or GSCM)[293], and LCA[294] have gained much in importance.[295]

A SC goes beyond the conventional value chain of a company and includes all parties which are involved in upstream and downstream flows.[296] "The supply chain includes not only the manufacturer and suppliers, but also the transporters, warehouses, retailers, and even customers themselves. [...] [It] includes all functions involved in [...] filing a customer request."[297] A GSC is a lean SC which produces less or zero

[293] ESCM and GSCM will be used as synonyms.

[294] The 'A' can be used either for 'assessment' or 'analysis'; see Tyteca (1996), p. 285.

[295] See Wang (2013), p. 3117; Hagelaar (2002), p. 399f; Nikbakhsh (2009), p. 196.

[296] See Chopra (2007), p. 3; Mentzer (2001), p. 4.

[297] Chopra (2007), p. 3.

waste, including the close work with and support of suppliers to improve their environmental impact.[298]

The traditional SCM is not greatly concerned about what happens with the product after its delivery and about any environmental impacts it may have.[299] Yet, environmental aspects have become of great interest for SC managers and have led to the development of GSCM that aims to reduce all environmental impacts through the whole SC going beyond the own organizational boundaries, even crossing borders and gaining global dimension, but also increasing productivity and profits at the same time. It can be defined as "'inte-grating environmental thinking into supply-chain management, including product design, mate-rial sourcing and selection, manufacturing processes, delivery of the final product to the consumers as well as end-of-life management of the product after its useful life'"[300]. This "supply chain environmental care strategy"[301] can be operationalized by implementing the respective EPIs. In order to be effective, ESCM requires comprehensive environmental information about all stages of a product's life.[302]

LCA allows exactly for this, by providing environmental information and evaluating direct and indirect environmental burdens, by making use of far-reaching analyses throughout the whole life-cycle, concerning products, processes or activities, and therefore is known as an important instrument of ESCM.[303] It assesses all inputs and outputs, their associated impact and emissions, and offers improvement options for environmental issues.[304] This methodology has been used in management accounting as 'life-cycle costing'.[305]

LCA might be used as basis for acquiring information that is necessary to become better able to select and implement the EPIs, although both may also represent two distinct tools.[306] Since LCA is still in its early stages, very comprehensive, time-

298 See Vijayvargy (2014), p. 26; Lu (2007), p. 4317.

299 See Nikbakhsh (2009), p. 197; Vijayvargy (2014), pp. 25, 27; Wolters (2002), p. 232, incl. following sentence.

300 Srivastava (2007), p. 54f.

301 Hagelaar (2002), p. 400, incl. whole sentence.

302 See Hagelaar (2002), p. 402; Nikbakhsh (2009), pp. 195, 197.

303 See Hagelaar (2002), pp. 400, 403; Rainey (2010), pp. 63, 552; Schaltegger (2000a), p. 241f; Nikbakhsh (2009), p. 201.

304 See Nikbakhsh (2009), p. 201; Rainey (2010), p. 63.

305 See Schaltegger (2000a), p. 240; for further information see Schaltegger (2000a), p. 124f.

306 See Benedetto (2009), p. 901; Lundin (2002), p. 145f; Schaltegger (2000a), p. 305; Olsthoorn (2001), p. 456; Hermann (2007), pp. 1787f, 1790.

consuming and expensive,[307] it is certainly not a must. However, it is offered here as an option for retrieving comprehensive environmental information[308].

6.2 Porter's Value Chain and its Adaptation

The general purpose of Porter's Value Chain[309] is its usage as a strategic tool for analysing a firm's possible sources of competitive advantage, in order to develop new ones or exploit existing ones.[310] It can also be used for structuring a firm into its different organizational units, including certain activities.[311] In this case, the use of the value chain draws on the systematic structuring of the indicators.

As section 6.1 has shown, there is a need for a holistic view when concerning oneself with the firm's overall environmental performance.[312] It is essential to consider not only the own value chain but to regard it in the context of a value chain system, including the suppliers and the channel value chains.[313] One may ask why the indicators will not be developed along the SC, the issue is that there is no clear definition or at least many firms understand it differently.[314] Some firms only include their (immediate) suppliers or only customers, some focus on internal activities. Consequently, the well-defined value chain of Porter was preferred.

It is not only value chains that differ among industries, but also within the same industry, but according to Porter all value chains have the same nine generic categories.[315] These categories are divided into two groups, primary activities and support activities.[316] The primary activities can be subdivided into five categories:[317]

- *Inbound logistics* includes all activities that are covering the handling of inbound materials and the contact to suppliers.
- *Operations* comprise all activities for transforming the input into an end product.
- *Outbound logistics* is mainly responsible for distribution as well as warehousing and product handling.

---

307 See Rainey (2010), p. 553; Lundin (2002), p. 146; Hermann (2007), p. 1787.

308 See Hermann (2007), pp. 1787–1789; Schaltegger (2000a), pp. 240–242.

309 See Porter (1985), pp. 33–61.

310 See Porter (1985), pp. 33, 59.

311 See Porter (1985), p. 59.

312 See Yang (2011), pp. 1180, 1182.

313 See Porter (1985), p. 34f; see Appendix 6.

314 See Charter (2001), pp. 3.2.2, 3.3.2, incl. following sentence.

315 See Porter (1985), p. 34; if a firm has outsourced some of its value chain activities it should still consider all indicators as the outsourced activities' environmental impact also affects the focal company.

316 See Porter (1985), p. 38; for visualization see Appendix 7.

317 See Porter (1985), p. 39f, incl. following key points.

- *Marketing and sales* is associated with creating a solid customer base by offering the means for buying the products and promoting them.
- *Service* includes all activities for customer support.

Although some activities are more meaningful to specific industries, all of them are apparent to a certain extent in each value chain.[318]

The second type of activities, are those supporting the primary activities as well as themselves.[319] The four generic subgroups are the following:

- *Technology development* includes product design, testing and process engineering, thus all activities associated with R&D and beyond.
- *Procurement* is concerned with the evaluation and monitoring of suppliers and with the purchase of inputs.
- *Human Resource Management* comprises all activities associated with personnel.
- *Firm infrastructure* covers all remaining support activities such as quality management, finance and accounting, legal affairs and general management.

When developing environmental indicators based on the value chain, it will be useful to modify it slightly for this purpose. Section 6.4[320] will orientate on the following value chain model in figure 1 which has emphases to certain activities. This model is still mainly based on Porter's value chain but it is including some small adaptations.[321]

318 See Porter (1985), p. 40.

319 See Porter (1985), pp. 40–43, incl. following key points.

320 The selection of the topics addressed by the indicators in section 6.4 is also supported by Epstein (2008b), pp. 90–92.

321 Few modifications (brackets) are based on Yang (2011), p. 1182; for original see Appendix 7.

Figure 1: Modification of Porter's Value Chain

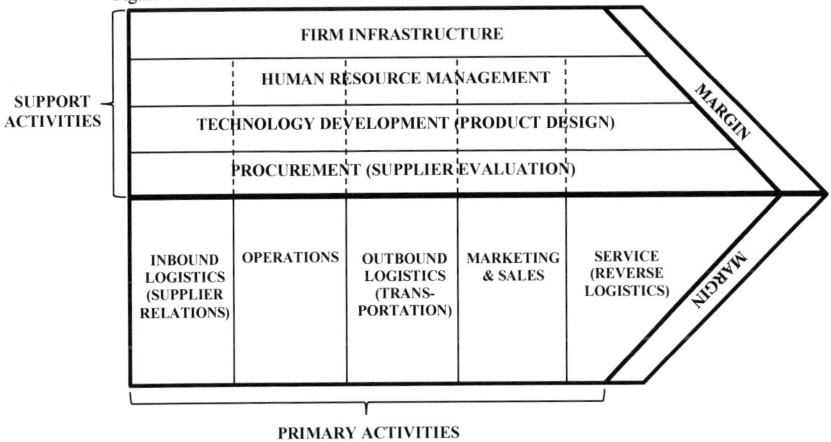

Source: Own representation based on Porter (1985), p. 37; Yang (2011), p. 1182.

The activities which are stated in brackets in figure 1 are especially important for the respective value chain unit. Procurement and inbound logistics with special reference to suppliers, as well as operations, will allow for an in-depth analysis for the environmental indicators. If it is possible to green the whole value chain, it will likely increase the economic performance and competitiveness of the firm.[322]

6.3 Structure and Categorization

In the following, the indicators will be divided into their functional units according to the different value chain activities as discussed in section 6.2 with special reference to suppliers. In some units there will be a further classification into the environmental media, referring mainly to the following affected environmental areas: energy, water and material, presenting the input, as well as waste and emissions, being the results of the different inputs.[323] Some value chain units will present indicators (in monetary units) which may be perceived as economic indicators, as they present environmental costs, investment and savings. Yet, as they are directly linked to environmental aspects, they can also be seen as environmental indicators (EMIs), which is the case in this paper.[324] They have been assumed to be an important complement to EPIs, and

[322] See Rao (2005), pp. 910, 912.

[323] See Müller (2011), p. 59; Schaltegger (1995), p. 34; Schaltegger (2000a), p. 299; Schröder (2014), p. 259; Ditz (1997), p. 6f; Noci (1997), p. 105; for diagrams with all processes, divisions and resources going in and out see Appendix 8, 9 and 10.

[324] See also e.g. GRI (2013), p. 35; DIN (2013), pp. 22, 55-57, incl. following sentence.

both should be tracked to be able to see the financial strain and all impacts the environmental commitment has.

In order to facilitate the entry into this topic and to allow later for comparisons between companies; generic indicators will be presented which may apply to all companies and covering the key environmental media.[325] They therefore serve as starting point and guidance for further enhancement. Nevertheless, it depends on the decision-maker which indicators to choose as they may have different effects on different firms, being more useful for one than for another depending on the specific situation.[326] Thus, the following should not be taken as a guaranteed route to success, but may give a good impression of how it can be when trying to make the whole value chain, including suppliers, environmentally sustainable. Yet, in order to green the company, it may be reasonable to start internally, in the "*Sphere of Control*"[327] before trying to green the whole SC, including the suppliers.

6.4 Specific Divisional Environmental Indicators[328]

6.4.1 Important Remarks

There are several important considerations when providing indicators for the different value chain functions. In order to apply indicators correctly some information has to be provided by the company.[329] The indicators should be described sufficiently with their corresponding units of measurement (such as kg, numbers, euros, percent, joules etc.)[330], the type of measurement (absolute or relative), the period they comprise (e.g. month, quarter, (fiscal) year), the geographical area they cover and other organizational boundaries (overall firm, facilities, departments, cost centres, products or product lines, materials), any uncertainties concerning the results and they should be reviewed and adjusted if needed. If referring to the respective addressees, indicators for lower levels such as divisions or cost centres should be assessed more regularly in order to react quickly to changes.[331] To be able to achieve viable information from

325 See Schaltegger (2000a), p. 299; Jasch (2002), p. 48f; Probst (2006), p. 20f; Veleva (2001), pp. 524, 527, incl. following sentence.

326 See Krajnc (2003), p. 287. For industry specific indicators see e.g. Hesse (2010).

327 Schröder (2014), p. 262, incl. whole sentence.

328 The delimitation of each value chain unit is somewhat fuzzy since some indicators may apply for several value chain units. Yet, the stated indicators for each unit are as specific as possible.

329 See DIN (2013), p. 31; Krajnc (2003), p. 282; Jasch (2000), p. 81f; Veleva (2001), p. 522; IFAC (2005), p. 32, incl. following sentence.

330 Also often measured in multiples, such as GJ or tons. They should be applied consistently in uniform units, which will be based on the International System of Units (SI units) in this paper in order to allow for an international applicability; see BIPM (2006).

331 See Jasch (2009), p. 55.

indicators, both absolute and relative indicators should be applied.[332] It gives a clear picture of the overall situation, as the relative indicator (e.g. relative consumption) may have diminished during the last years, while the absolute indicator (e.g. total consumption) may have still risen due to an increase in production.[333] With regards to environmental impacts, absolute indicators are most significant, since they clearly show the total consumption and environmental impact.[334] In order to make the absolute indicators truly useful they have to be assessed against a benchmark, this means a certain baseline, using e.g. official limit values.[335] However, relative or normalized indicators are crucial indicators as not only the environmental performance but also the organization changes in size, production and service output over the years.[336] Moreover, they allow for direct comparisons as they relate to significant denominators and show improvements in efficiency.[337] As already mentioned, relative indicators are insufficient in order to present the real development of environmental impact, since an increasing output may offset an increasing environmental impact.[338] The critical aspect of relative indicators is the metric of their denominator "that must be precisely defined and logically related to the basic indicator."[339] There is a large amount of suitable or less suitable denominators.[340] The *production output* is a very important one.[341] "The quantity of products produced and sold is the most useful denominator, preferably measured in kg, sometimes in volume or number. If physical data is not available, turnover in monetary terms is the second best choice."[342] However, if it is used for comparisons, the assumption of similar products has to be made.[343] As an alternative the *turnover* or *cost of production* might be a very good choice as it is directly related to the production output and in the beginning volume data may not be available but expenditures are.[344] Yet, if the firm's production and products, as well as their environmental impacts differ a lot, it is not very useful. *Net sales* is another possible denominator, adapted from turnover by sales discounts, returns and allowances, but like turnover it might be overestimated.[345] One has to recognize that monetary sales

332 See Jasch (2009), p. 56; Veleva (2001), p. 527; Müller (2011), p. 61.
333 See Veleva (2001), p. 527.
334 See IFAC (2005), p. 32; Jasch (2009), p. 56.
335 See Müller (2011), p. 61; OECD (2011), p. 39.
336 See IFAC (2005), p. 32.
337 See Jasch (2009), p. 56.
338 See Jasch (2009), p. 56; Müller (2011), p. 61.
339 Jasch (2009), p. 58, incl. whole sentence.
340 The following is only a limited selection.
341 See Jasch (2009), pp. 58, 60; Jasch (2000), p. 81.
342 Jasch (2009), p. 60; for support of this quotation see Jasch (2009), p. 58; Jasch (2000), p. 81, although there is often referred to as tons instead of kilograms.
343 See Linke (2013), p. 558; Jasch (2009), p. 58.
344 See Jasch (2009), pp. 58, 60, incl. following sentence.
345 See Jasch (2009), p. 60; Olsthoorn (2001), p. 457f, incl. following sentences.

units do not change equally to production volumes, as they are affected by price, currency exchange rate, demand and stock changes. The *value added* equals net sales subtracted by costs of goods and services purchased. However, as this indicator is not mandatory or clearly defined by all, it may not be applied due to comparison differences between worldwide firms. *EBIT* is a further possible denominator, as well as *net profit after tax* but the latter is rather unsuitable as it is easily distorted by different influences.[346] Another possibility is the number of *full-time equivalent* (*FTE*) or *person-hours* worked but it might "depreciate automated manufacturing"[347] and there are issues concerning different labour intensities between different sectors.[348] Another one which could provide important information but which is difficult to determine is a product's *durability*.[349] Sometimes it may also be necessary to make use of weighting indicators, converted in an importance scale.[350]

6.4.2 Inbound Logistics and Procurement with special reference to DC Suppliers

6.4.2.1 Supplier Management and Environmentally Conscious Purchasing

As it has been stated in chapter 3, DCs are still shaped by uncertainty and specific country risks which may put pressure on the buyer-supplier relationship.[351] There are four principal supplier management tasks, "supplier selection and evaluation, supplier monitoring, supplier development, and supplier integration"[352]. The indicators in subsection 6.4.2.3 are presented for the use of the first two aspects while the latter ones may be very important for the DC supplier management and should not be neglected but won't be addressed in this paper.[353] Nevertheless, also for supplier selection, evaluation and monitoring one should bear in mind the more difficult environments DCs are doing business in, by having limited means and knowledge, as well as facing more issues than developed country suppliers. The conclusions drawn from the assessment of the environmental indicators should also be communicated to the suppliers, and improvement should be supported through training offerings, personnel support, provision of expertise, as well as investment and transfer of resources, instead of only doing strict evaluations and eliminating suppliers. Supplier environmental management has taken different forms, such as questionnaires, requirements for ISO standards and other certifications, direct site assessments, or more profound

346 See Jasch (2009), p. 61.

347 Linke (2013), p. 559.

348 See GRI (2013), p. 158; Linke (2013), p. 558f; Olsthoorn (2001), p. 457f.

349 See Linke (2013), p. 558f.

350 See Linke (2013), pp. 556, 559, 562; Jasch (2000), p. 81.

351 See Akamp (2013), p. 55f.

352 Akamp (2013), p. 55, incl. whole sentence.

353 See Akamp (2013), p. 55f; Rao (2002), p. 636; Khan (2014), pp. 1212, 1220; Trapp (2014), p. 3; Weber (2014), p. 134f, incl. following sentences.

approaches like partnering or mentoring.[354] It should not be forgotten that creating an environmentally friendly international SC is certainly not an easy procedure but a difficult journey.[355] Despite the fact that many firms all over the world put more and more emphasis on sustainability and are aware of its relevance and significance.[356]

However, if this challenge is abandoned, the SCM and procurement will have negative impact on the firm's direct and indirect pollution and waste, as well as on the perceived value to customers, and therefore affects the environmental and economic performance of the firm.[357] If it is adequately addressed negative environmental impacts can be stopped at the source.[358] Suppliers are no longer only required to produce most cheaply, offering the highest quality in the shortest time, but now must also take environmental aspects seriously.[359] The introduction of the ISO 14000 certification standards as well as environmental requirements from the Environmental Protection Agency (EPA) sped up this development.[360] As environmental regulations are particularly strict in European countries, a German company should take the environmental management of suppliers very seriously.

Suppliers are critical since their supplied materials and products, and their production processes have a significant impact on the buyer company's environmental situation, business and image, as firms are nowadays blamed for the mistakes done by suppliers, and suppliers are assumed to be of main importance for the whole SC to become sustainable.[361] Particularly, since DCs have grown to critical suppliers, also as manufacturing hubs. Therefore, environmental issues become of great concern for the purchasing unit, also known as environmentally conscious purchasing (ECP).[362] There is confirmation that ECP benefits the company's performance.[363] Thus, EMA, and here especially environmental indicators, could play an important role in ESCM.[364]

---

354 See Rao (2002), p. 633f.

355 See Lahme (2014), p. 155.

356 See Ahi (2015b), p. 2882; chapter 3; section 2.2.

357 See Handfield (2002), p. 71; Govindan (2014), p. 212.

358 See Rao (2008), p. 24.

359 See Handfield (2002), p. 71; Lee (2009), p. 7917f; Ellram (1990), p. 8; Kumar (2014), p. 109.

360 See Handfield (2002), p. 71, incl. following sentence.

361 See Rao (2008), pp. 24f, 68; Mani (2014), pp. 99, 102; Noci (1997), p. 104; Büyüközkan (2011), p. 164; Rao (2002), pp. 632, 637; Rao (2014), p. 60; Rao (2005), p. 903; Schröder (2014), p. 265; Brown (1996), p. 111, incl. following sentence.

362 See Handfield (2002), p. 72f.

363 See Carter (2000), p. 224.

364 See Wolters (2002), p. 236.

6.4.2.2 Conceptual Delimitation and Research Approach

The aim of this subsection is not to give a detailed methodology, approach or model for green supplier selection and evaluation as is already done in a number of research papers, with the help of e.g. applied mathematical modelling, Analytical Hierarchy Process (AHP), analytical network process, grey relational analysis, (novel/hybrid) fuzzy multicriteria approach/decision framework, fuzzy TOPSIS, fuzzy Data Envelopment Analysis (DEA), grey system theory, (neighbourhood) rough set methodology, integrated evaluation methodology, performance measurement reduction methodology, maturity stage model, or the Ishikawa diagram.[365] This subsection aims to give a coherent choice of indicators for selecting, evaluating and monitoring DC suppliers[366] according to their environmental sustainability.

In this case, with reference to suppliers being located in DCs the corresponding indicators have been primarily chosen by considering that they should be particularly simple and easily understood and assessed; applicable, measurable and relevant.[367] The selection of the following environmental indicators for DC suppliers will be mainly based on the indicators used in the practical 'supply chain environmental sustainability scorecard' of Procter & Gamble, which has already led to improvement and greater collaboration in their supply network, and is proven in its applicability, profitability, relevance and positive impact on the environment; on the environmental indicators proposed by the GRI; and on the comprehensive research of Ahi and Searcy (which made use of a content analysis comprising 445 articles and led to 2555 metrics which were ordered according to their appearance).[368] These sources have been supported by findings from the sources named in footnote [365] (some making use of a Delphi study, interviewing decision makers or experts to weight the respective EPIs for their importance)[369] in order to get an in-depth knowledge of the available and utilized environmental indicators for suppliers with a few sources concentrating on DC suppliers[370]. The analysis of Ahi and Searcy has shown that consistent indicators are still rare and many different metrics exist for the same issue.[371] Thus, there is a need for consolidation and generic indicators to enable comparisons between firms.

---

365 For more details and different SC models and supplier selection/evaluation models see i.a. Ahi (2015a); Ahi (2015b); Awasthi (2010); Azadi (2015); Bai (2010); Bai (2012); Bask (2011); Büyüközkan (2011); Dobos (2014); Enarsson (1998); Erol (2011); Govindan (2013); Handfield (2002); Hashemi (2015); Hsu (2009); Huang (2007); Humphreys (2003); Kumar (2014); Lu (2007); Noci (1997); Öztürk (2014); Saen (2009); Shen (2013); Tuzkaya (2009); Rao (2008), pp. 47–53.

366 This does not mean that the indicators cannot be applied to developed country suppliers.

367 See subsection 5.4.3.

368 See P&G (URL1); P&G (URL2); GRI (2013); Ahi (2015a), p. 360. For P&G's supplier scorecard see P&G (2014): Excel sheet 3. P&G's indicators are mainly based on GRI.

369 See Awasthi (2010), p. 373; Handfield (2002), pp. 76–78.

370 See Rao (2002); Akamp (2013).

371 See Ahi (2015a), pp. 366, 371, 374, incl. following sentence.

6.4.2.3 Selecting and Monitoring Suppliers with the Help of Indicators[372]

Conventionally, suppliers have been evaluated according to i.a. their prices, quality, flexibility and economic efficiency.[373] As the environmental situation has become alarming, firms have started to consider environmental aspects in their supplier evaluation as well. In terms of the selection, suppliers are evaluated on whether they can become a member of the SC or not, some firms are only working with suppliers who fulfil specific requirements.[374] Since many firms may already have DC suppliers but no environmental monitoring in use, the following indicators may be primarily used for initial evaluation and monitoring actions.

The indicators presented below are grouped in their different environmental categories, as well as in core and optional measures. The core measures present generic indicators which probably can be applied by all companies, while for the optional indicators it may not be possible to collect the required data as the suppliers are still from DCs or the indicators are not as relevant. Some measures in the end of this subsection will also be qualitative, but can be used to complete the environmental evaluation.

Core Measures[375]

The following indicators are assumed to be most relevant to be tracked and assessed. Regardless of the different kinds of indicators, the suppliers have to report which scope the indicators are based on, this means whether the supplier offers product (directly related to the buyer firm's purchased materials or products), site or corporate information.[376]

ENERGY[377]

Energy is one of the main corporate environmental issues, which also became clear due to its great presence in literature.[378] A company's overall environmental footprint

---

372 Some indicators will be explained in more detail in subsection 6.4.6. This subsection only deals with the part of inbound logistics and procurement concerning suppliers. Some aspects which are not directly dealing with suppliers will be handled in subsection 6.4.2.4.

373 See Ellram (1990), p. 8; Weber (1991), pp. 2, 4, 12, 14; Lee (2009), p. 7918; Enarsson (1998), p. 9; Kumar (2014), p. 109; Awasthi (2010), p. 370; Öztürk (2014), p. 133, incl. following sentence.

374 See Rao (2002), p. 632; Saen (2009), p. 144; Handfield (2002), p. 74; Rao (2010), p. 2; Brown (1996), p. 117.

375 The formulas, terms and explanations of the following indicators are mainly derived and taken from ('derived': the concrete formula has been created or adapted by the author; 'taken': the formula has been adopted) P&G (2014): Excel sheet 2. The direction of the arrows next to the formula shows whether the indicator shall be reduced or increased over time.

376 See P&G (2014): Excel sheets 2&3.

377 The following indicators ('indicator' always addresses the indicator formula as well) are mainly derived from P&G (2014): Excel sheet 2; GRI (2013), pp. 89–93.

378 See Ahi (2015a), pp. 360, 366; Enarsson (1998), p. 13; Erol (2011), p. 1093; GRI (2013), pp. 89–93; Lu (2007), p. 4324f; Noci (1997), p. 109.

is influenced by its energy consumption,[379] and therefore also important for the buyer company, in order to gain a better understanding of how much energy the purchased materials or products have already consumed. Electricity may present the main or only important energy consumption of some supplier companies, while heat and steam may be for others. Energy from fuel will be measured separately to electric energy.[380] This allows the energy sources the suppliers are using predominately to be highlighted.

I. (Electric) Energy Consumption

Absolute Indicator:

$$(1)\ Total\ Direct\ Net\ Energy\ Usage\ (GJ) = Purchased + Self\text{-}generated - Sold \quad \downarrow$$

Relative Indicator:

$$(2)\ (Electric)\ Energy\ Intensity = \frac{Total\ Direct\ Net\ Energy\ Usage\ (GJ)}{Unit\ of\ Output} \quad \downarrow$$

The absolute indicator (1) consists of the total direct net energy usage from electricity measured in gigajoules which the supplier is consuming for all kinds of operations, such as production, office lighting, heating or cooling.[381] In order not to double-count the self-generated electricity, it is only counted once for fuel consumption, and as the sold energy will not be used by the supplier himself it is subtracted.[382] This indicator should certainly be reduced over time. The relative indicator (2)[383] shows the intensity ratio by dividing the total energy usage (numerator) by an organization-specific unit of output (denominator).[384] The units of output may be[385] metric tons (mt)[386], cubic metres (m^3) or gigajoules (GJ) (production turnover), square metres (m^2) (floor space), number of full-time equivalents (FTE), monetary units such as revenue (sales) or value

379 See GRI (2013), p. 89, incl. following sentence.

380 See P&G (2014): Excel sheet 2.

381 See P&G (2014): Excel sheet 2; GRI (2013), p. 89.

382 See GRI (2013), p. 89. Energy sources will be further explained in subsection 6.4.6.

383 'The less the better' applies for all relative indicators which show the impact per unit; see Schwarz (2002), p. 59.

384 See GRI (2013), p. 93; P&G (2014): Excel sheets 2&3, incl. following sentences.
It would be advantageous if all suppliers make use of the same denominator for the same indicators but especially as it refers to DC suppliers some may not have all data available and the choice should be up to them. Yet, the selected unit should remain the same over the years.

385 This should be adapted to the specific organizational situation of each company.

386 In general the term ton is used but in English speaking countries the term metric ton is usually used. Therefore, in an international context with regard to suppliers, 'mt' should be used to be clear about the concrete meaning to all; see BIPM (2006), p. 124.

added[387]. With the help of this intensity ratio, energy consumption becomes comparable to other years and firms, showing progress,[388] as long as the same unit of output is applied. The suppliers may find information through invoices, measurements, calculations or estimations, or through the units directly taken from meters which have to be translated into the reported units.[389]

Concerning fuel consumption the following applies. Fuel sources consist of "fuel for combustion in boilers, furnaces, heaters, turbines, flares, incinerators, generators and vehicles, which are owned or controlled by the organization."[390] The calculation of these indicators is equal to the calculation for electric energy.

II. (Fuel) Energy Consumption

Absolute Indicator:

$$(1)\ Total\ Direct\ Net\ Energy\ Usage\ (GJ) = Purchased + Self\text{-}generated - Sold \quad \downarrow$$

Relative Indicator:

$$(2)\ (Fuel)\ Energy\ Intensity = \frac{Total\ Direct\ Net\ Energy\ Usage\ (GJ)}{Unit\ of\ Output} \quad \downarrow$$

The total direct net energy usage (1) is the sum of all kinds of fuels (e.g., fuel oil, natural gas) for all possible needs (e.g. manufacturing, food production, heating, etc.).[391] In the case of suppliers only the energy consumed within the firm gets measured, as a larger scope may exceed their competences.

WATER [392]

As it could be seen in chapter 4, water is scarce, and presents a critical resource. In order to become an environmentally sustainable company, the supplier has to show an appropriate handling of water.

[387] See Azapagic (2000), p. 246; Jasch (2009), p. 60; value added may be difficult in terms of DC suppliers as it may not be calculated by them; Jasch (2009), p. 60.

[388] See P&G (2014): Excel sheet 2.

[389] See GRI (2013), p. 90. For converting different energy units into joules see Appendix 11 and 12.

[390] GRI (2013), p. 89, incl. whole sentence.

[391] See P&G (2014): Excel sheet 2; GRI (2013), p. 89, incl. following sentence.

[392] The following indicators are mainly derived from P&G (2014): Excel sheet 2; GRI (2013), pp. 97-99, 122.

III. (Input / Withdrawal) Water Usage

Absolute Indicator:

(1) $Total\ Water\ Withdrawal\ (m^3)$ ↓

Relative Indicator:

(2) $Water\ Intensity = \frac{Total\ Water\ Withdrawal\ (m^3)}{Unit\ of\ Output}$ ↓

The absolute indicator (1) shows the total quantity of water in cubic metres (m^3) that is coming into the supplier's site from any sources, directly or through intermediaries, such as surface water, ground water, waste water from other organizations, water bought from local water supplies or pumped from the ocean or lake, or from the abstraction of cooling water.[393] Rainwater is excluded from this indicator, "unless collected and used in process area"[394]. The water intensity (2) is similarly calculated to the energy intensity, showing the water usage per unit of output that makes it again comparable.

The total volume of water discharged to the natural environment has great ecological impact on its surroundings, concerning the total amount and the effluent load of chemicals and nutrients.[395]

IV. (Output / Discharge) Water Usage / Waste Water

Absolute Indicator:

(1) $Total\ Water\ Discharge\ (m^3)$ ↓

Relative Indicator:

(2) $Waste\ Water\ Intensity = \frac{Total\ Water\ Discharge\ (m^3)}{Unit\ of\ Output}$ ↓

In order to create the absolute indicator (1), the supplier firm has to provide data about the total amount of water leaving the site, not including rainwater or domestic sewage. The data can be received through flow meters, or if not available through estimates, when the firm subtracts the volume of water consumed from the volume withdrawn. The relative indicator (2) provides a relation to units of output for better comparisons.

393 See P&G (2014): Excel sheet 2; GRI (2013), p. 97; the total water intake is measured, whether it is directly consumed by the product, for production or sanitary purposes.

394 P&G (2014): Excel sheet 2, incl. whole sentence.

395 See GRI (2013), p. 122; P&G (2014): Excel sheet 2, incl. following sentences.

MATERIAL[396]

As suppliers are delivering or also consuming raw materials (minerals, ores, etc.)[397] or others (e.g. in food production: coffee beans, grains, etc., or packaging materials[398]) coming from their own suppliers or directly, they also contribute to raw material extraction and their decline, and so does the buyer firm in the end.

V. (Raw) Material Consumption

Absolute Indicator:

(1) $Total\ Material\ Consumption\ (mt)$ ↓

Relative Indicator:

(2) $Material\ Intensity = \frac{Total\ Raw\ Material\ Consumption\ (mt)}{Unit\ of\ Output\ (Production)}$ ↓

The total material consumption (1)[399] includes all kinds of (raw) material purchased, used, transferred or consumed by the supplier measured in metric tons.[400] The relative indicator (2) shall provide a measure to show the efficiency of the supplier. As material is directly linked to production the denominator shall consist of a production unit, such as metric tons, gigajoules, cubic metres, etc.,[401] therefore presenting how much material has been used in order to produce one unit of output. The goal of the supplier should certainly be to reduce this metric. Thus, either reducing the input while having the same output[402] or staying with the input and increasing the output. Eventually, the overall goal should be to reduce the input and increase the valuable output and replace all environmental harmful or not recyclable or renewable (raw) materials with environmentally friendly materials. Thus, it may be also very interesting to measure the share of environmentally friendly material consumption,

396 The following indicators are derived from the indicators given by P&G (P&G does not include these indicators) but the relevance and importance could be concluded by the appearance in literature see GRI (2013), p. 86f; Ahi (2015a), p. 371; Awasthi (2010), pp. 371, 374; especially Huang (2007), p. 521; and many authors referring to further authors making use of material indicators: Shen (2013), p. 175; Hashemi (2015), p. 179f; Govindan (2013), p. 347; Öztürk (2014), p. 134; Bai (2010), p. 254; Azadi (2015), pp. 275, 278.

397 See GRI (2013), p. 86.

398 Ibid.

399 The purpose of this indicator depends on the kind of supplier. Some suppliers may deliver the raw materials directly; others may supply semi-products which have already consumed raw materials. As the buyer firm will calculate its own material consumption indicator, this supplier indicator may serve as direct information source or as an additional indicator if it shows the raw material consumption for semi-products for instance. Moreover, it shows the handling of suppliers with material, and therefore their environmental commitment and efficiency.

400 See GRI (2013), p. 86f, incl. following sentence.

401 See GRI (2013), p. 93.

402 See Goldmann (2001), p. 598.

including renewable or recyclable/recycled materials.[403] This might be assessed in a further step as it requires a very precise monitoring of the suppliers production processes.[404]

WASTE[405]

If the total weight of waste is tracked periodically, the efforts in waste reduction the supplier has made can be disclosed.[406] It may also indicate the improvement in efficiency and productivity. The waste might be categorized into hazardous and non-hazardous waste.

VI. Hazardous Waste (Disposal)

Absolute Indicator:

$$(1)\ Total\ Weight\ of\ Hazardous\ Waste\ (mt) \downarrow$$

Relative Indicator:

$$(2)\ Hazardous\ Waste\ Intensity = \frac{Total\ Weight\ of\ Waste\ (mt)}{Unit\ of\ Output} \downarrow$$

The total weight of hazardous waste (1) consists of waste which is incinerated or landfilled, and has not been subject to energy recovery, measured in metric tons.[407] The waste that is reused, recycled or recovered is excluded from this indicator and will be offered as an optional indicator (recycling). 'Hazardous' waste is defined by national regulations and differs between countries. Possible hazardous waste comprises "solvents, flammable/hazardous raw materials, chemical waste, batteries, acids (low pH liquids), strong bases or caustic (high pH liquids), etc."[408] The waste intensity (2) presents again an indicator useful for comparisons.

[403] See Ahi (2015a), pp. 364, 369, 371f; Awasthi (2010), p. 374; GRI (2013), p. 87.
[404] See Rao (2008), p. 24.
[405] The following indicators are mainly derived from P&G (2014): Excel sheet 2; GRI (2013), p. 123.
[406] See GRI (2013), p. 123, incl. following sentences.
[407] See P&G (2014): Excel sheet 2, incl. following sentences.
[408] P&G (2014): Excel sheet 2.

VII. Non-Hazardous Waste (Disposal)

Absolute Indicator:

(1) $Total\ Weight\ of\ Non\text{-}Hazardous\ Waste\ (mt)$ ↓

Relative Indicator:

(2) $Non\text{-}Hazardous\ Waste\ Intensity = \frac{Total\ Weight\ of\ Waste\ (mt)}{Unit\ of\ Output}$ ↓

The total weight of non-hazardous waste (1) comprises all other kinds of waste that are landfilled or incinerated but not recovered.[409] The relevant data exposes the amount of solid waste but also liquids which are stored in solid containers. The weight, if not directly available, may be detected by estimation, making use of information of waste density and volume, mass balances or corresponding data.[410] The relative indicator (2) equals the previous one.

EMISSIONS[411]

As emissions also play a significant role in contributing negatively to the natural environment,[412] the respective measures should not be neglected by accounting. This also includes emissions on behalf of the suppliers that in turn, indirectly affect the buyer firm's own environmental footprint, and here especially its carbon footprint.[413] "The term 'carbon footprint' is commonly used to describe the carbon dioxide (CO2) and other GHG emissions for which an […] organisation is responsible."[414] Through measuring the supplier's GHG emissions its environmental efficiency becomes further visible. These emissions are based on the regulations of the WRI and WBCSD 'GHG Protocol Corporate Accounting and Reporting Standard', also known as the GHG Protocol.[415] It classifies the GHG emissions in three different Scopes – Scope 1, 2 and 3. These scopes define the boundaries in which the emissions prevail.[416]

---

409 See P&G (2014): Excel sheet 2, incl. following sentence.

410 See GRI (2013), p. 123.

411 The following indicators are derived from P&G (2014): Excel sheet 2; GRI (2013), pp. 107–115; WBCSD (2004), pp. 25–28; Jasch (2002), p. 48.

412 See chapter 4.

413 For carbon footprint see Kumar (2014), p. 109; Carbon Decisions (2010).

414 Kumar (2014), p. 112, incl. following sentence.

415 See GRI (2013), p. 105; WBCSD (2004), pp. 3, 25, incl. following sentences.

416 For a graphical explanation and differentiation of the three scopes see Appendix 13.

VIII. Direct Greenhouse Gas Emissions (Scope 1)

Absolute Indicator:

$$(1)\ Total\ Amount\ of\ Scope\ 1\ GHG\ Emissions\ (mt\ of\ CO2e)$$ ↓

Relative Indicator:

$$(2)\ GHG\ Emissions\ Intensity = \frac{Total\ Amount\ of\ GHG\ Emissions\ (mt\ of\ CO2e)}{Unit\ of\ Output}$$ ↓

The GHG emissions of Scope 1 (1) consist of all direct emissions, resulting "from sources (physical units or processes that release GHG into the atmosphere) that are owned or controlled by the organization."[417] The GHG emissions, measured in metric tons of CO_2 equivalent[418], comprise the following emissions: carbon dioxide (CO_2), methane (CH_4), nitrous oxide (N_2O), hydrofluorocarbons (HFCs), perfluorocarbons (PFCs), sulphur hexafluoride (SF_6) and nitrogen trifluoride (NF_3); according to the Kyoto Protocol.[419] CO_2 is especially critical, as it "is the main contributor to global warming"[420]. These emissions come primarily from corporate activities, such as the generation of electricity, resulting from the combustion of fuels; physical and chemical processes, mostly coming from the manufacturing or processing of materials and chemicals; transportation of all kinds of products, waste, materials and staff, through the combustion of fuels in corporate vehicles; and fugitive emissions, appearing from releases, consciously or unconsciously.[421] Emissions resulting from self-generated electricity, which will be sold, are not deducted from Scope 1. Suppliers may use their own, already existing tools for calculating the total amount of their GHG emissions.[422] If not, they can also make use of calculators offered online[423]. If the suppliers make use of sea transport, the respective GHG emissions should be added to the total amount. Assistance for the calculation is provided by a calculation tool of the GHG protocol[424]. Indicator (2) presents the GHG emissions intensity by showing the emissions per unit of output.[425] This allows for normalized environmental impact data. Concerning the tracking of this data, the suppliers should choose a "consistent

[417] GRI (2013), p. 107 based on WBCSD (2004), p. 25, incl. whole sentence.
[418] For further explanation see The Guardian (URL); OECD (URL1); Tjahjadi (1999), p. 89.
[419] See WBCSD (2004), p. 3; GRI (2013), p. 107.
[420] Färe (2004), p. 348, incl. whole sentence.
[421] See WBCSD (2004), pp. 25, 27, incl. following sentence.
[422] See P&G (2014): Excel sheet 2, incl. following sentences.
[423] See e.g. National Energy Foundation (URL).
[424] See The Greenhouse Gas Protocol (URL): 'GHG emissions from transport or mobile sources'.
[425] See GRI (2013), p. 115, incl. following sentence.

consolidation approach for emissions, and apply it to calculate the gross direct (Scope 1) GHG emissions."[426] They should also decide for a base year.

Scope 2 comprises "electricity indirect GHG emissions"[427]. All emissions resulting "from the generation of purchased electricity"[428] used for own purposes (operations and equipment) will be tracked in this scope. These indirect electricity emissions present a major source of GHG emissions for numerous companies, and therefore also offer a good opportunity for great progress in reduction.

IX. Electricity Indirect Greenhouse Gas Emissions (Scope 2)

Absolute Indicator:

(1) $Total\ Amount\ of\ Scope\ 2\ GHG\ Emissions\ (mt\ of\ CO2e)$ ↓

Relative Indicator:

(2) $GHG\ Emissions\ Intensity = \frac{Total\ Amount\ of\ GHG\ Emissions\ (mt\ of\ CO2e)}{Unit\ of\ Output}$ ↓

Indicator (1) consists of gross indirect GHG emissions coming from the generation of acquired electricity and used for own consumption.[429] Other indirect emissions (Scope 3) will not be included. As for Scope 1, a base year and a consolidation approach are needed.[430] The firm should ask the suppliers to offer all information regarding standards, methodologies, and assumptions made for their calculations. As for the Scope 1 relative indicator (2), the same applies for this.

COSTS[431]

Environmental costs also have a great significance in supplier evaluation when looking at numerous researches.[432] Environmental spending by suppliers, either for the treatment of waste and pollution or for improving the firm's environmental situation[433] provides some evidence towards the suppliers' commitment, and supplements the previous environmental indicators which have shown physical improvements. Yet, making use of indicators for environmental expenditures alone would not be very significant as it does not directly indicate the environmental performance, especially as

426 GRI (2013), p. 108, incl. whole sentence and following sentence.

427 WBCSD (2004), pp. 25, 27, incl. whole sentence and following sentences.

428 GRI (2000-2011), p. 27.

429 See WBCSD (2004), p. 27; GRI (2013), p. 110, incl. following sentence.

430 See GRI (2013), p. 110, incl. following sentence.

431 The first two indicators are taken from Humphreys (2003), pp. 144f, 148.

432 See Ahi (2015a), pp. 366f, 369, 371; Awasthi (2010), p. 371; Azadi (2015), p. 275; Humphreys (2003), pp. 144f, 148; Noci (1997), p. 109f; Tuzkaya (2009), p. 486.

433 See Humphreys (2003), pp. 144f, 148.

monetary metrics can be easily distorted.[434] As the total amounts spent for the environment might be little compared to the amounts of developed county suppliers, the relative indicator does only give a concrete explanation in this case.

X. Environmental Costs

Absolute Indicators:

$$(1)\ Total\ (Pollutant\ Effects)\ Environmental\ Costs\ (EUR)$$
$$= Solid\ Waste\ Disposal\ Costs$$
$$+\ Chemical\ Waste\ Disposal\ Costs$$
$$\downarrow \quad +\ Treatment\ Costs\ for\ Air\ Pollutant$$
$$+\ Water\ Waste\ Treatment\ Costs$$
$$+\ Costs\ for\ Energy\ Consumed$$

$$(2)\ Total\ (Improvement)\ Environmental\ Costs\ (EUR)$$
$$= Costs\ for\ Buying\ Environmental\ Friendly\ Material$$
$$\uparrow \quad +\ Buying\ Environmental\ Friendly\ Technology$$
$$+\ Redesign\ Products\ +\ Staff\ Training\ +\ Recycling$$

Relative Indicator:

$$(3)\ Relative\ Environmental\ Spending = \frac{Total\ Environmental\ Costs\ (EUR)}{Revenues\ (EUR)}\ x\ 100$$

The total environmental costs for both pollutant effects and improvement have to be calculated.[435] As the indicators are in monetary terms and created for a German company, the currency will be Euro[436] and has to be converted by its suppliers. Indicator (1) refers to costs spent for waste disposal and treatment of pollutants.[437] Indicator (2) encompasses all investments made in order to enhance the environmental performance. To get the total costs concerning pollutant effects (1), the following will be included.[438] "Solid waste disposal costs" include all relevant treatment and transportation costs concerning the transportation of solid waste disposal. "Chemical waste disposal costs" consist of all costs concerned with the disposal of chemical materials being toxic and harmful. "Treatment costs for air pollutant" involve all costs which are spent in order to decrease the concentration of pollutants in the air before

[434] See Müller (2011), p. 59.
[435] See Humphreys (2003), pp. 144f, 148.
[436] If a firm's accounting unit uses another currency, this metric should be adapted respectively.
[437] See Humphreys (2003), p. 144, incl. following sentence.
[438] See Humphreys (2003), p. 148, incl. following sentences and quotations.

being released. Costs for energy consumed are all costs for energy that was used for production and reveals whether a supplier works efficiently. The environmental costs for improvement (2) are composed of the following costs. The "costs for buying environmental friendly material", instead of any kind of virgin material. "Costs for buying new equipment", including new technology that helps to improve the environmental performance. Product redesign costs, resulting in products which are more easily disassembled or need less material. Costs to train staff to increase their expertise concerning environmental issues and clean technology. Finally, the costs which come from recycling the products in order to reduce the disposal of waste. If both environmental performance and environmental investment are measured, it shows the suppliers' efficiency in utilizing its environmental budget, as one will see the environmental improvements resulting from the investment.[439] The total amount of each kind of environmental cost serves as the numerator of indicator (3). This shows how much percent of revenue was spent for environmental issues and improvement actions, and makes it therefore comparable to other suppliers. These indicators may also show if the DC supplier needs monetary support in order to enhance its environmental performance.

All stated indicators are applicable to DC suppliers as they should be 'easily' and reliably determined, and will already provide comprehensive knowledge for the purchasing company about its current or potential future suppliers' environmental sustainability.

Optional Measures

The optional measures present indicators that supplement the core measures, and may further enhance the control and the provided data concerning the suppliers' environmental situation. However, these measures are only presented as optional as they may not be calculated[440] from all suppliers due to lacking means and knowledge or relevance. If it is possible to receive those, they should, but if not, it is no great issue as long as the core measures can be obtained. Yet, for future, it may be beneficial to support the suppliers in the improvement of their equipment and knowledge, in order to enable them to offer the following indicators.

---

[439] See GRI (2013), p. 135.

[440] See P&G (2014): Excel sheet 2.

Energy (Renewable Energy)

$$(1)\ Renewable\ Energy\ Consumption = Total\ Renewable\ Energy\ Used\ (GJ)\ \nearrow$$

$$(2)\ Renewable\ Energy\ Intensity = \frac{Total\ Renewable\ Energy\ Used\ (GJ)}{Unit\ of\ Output}\ \nearrow$$

$$(3)\ Relative\ Renewable\ Energy\ Consumption = \frac{Total\ Renewable\ Energy\ Used\ (GJ)}{Total\ Energy\ Consumption\ (GJ)}\ x\ 100\ \uparrow$$

Recycling

$$(4)\ Total\ Amount\ of\ Waste\ Material\ Recycled, Reused, Recovered\ (mt)\ \nearrow$$

$$(5)\ Recycling\ Rate = \frac{Total\ Amount\ of\ Recycled\ Waste\ Material\ (mt)}{Total\ Amount\ of\ Waste\ Material\ (mt)}\ x\ 100\ \uparrow$$

Emissions

$$(6)\ Other\ Indirect\ GHG\ Emissions\ (Scope\ 3) = Total\ Amount\ of\ Emissions\ \downarrow$$

$$(7)\ Other\ Indirect\ GHG\ Emissions\ Intensity\ (Scope\ 3) = \frac{Total\ Amount\ of\ Scope\ 3\ GHG\ Emissions\ (mt\ of\ CO2e)}{Unit\ of\ Output}\ \downarrow$$

Transportation

$$(8)\ Transportation\ Fuel\ Efficiency = \frac{Emissions\ of\ CO2\ (grams\ of\ CO2)}{Transport\ Unit\ (ton\text{-}km)}\ \downarrow$$

Compliance

$$(9)\ Fines\ \&\ Sanctions = Total\ Amount\ of\ Monetary\ Fines\ (EUR)\ \&\ Total\ Number\ of\ Non\text{-}monetary\ Sanctions\ \downarrow$$

The first possible additional indicator sphere refers to energy, in particular renewable energy.[441] This environmental aspect may seem rather important but it is only posed as optional, as it may be possible that especially DC suppliers do not make use of renewable energy yet. Therefore, it is only used for those where it applies. The total amount of renewable energy (1) is measured in gigajoules and comprises the "total energy used that comes from both biogenic fuels (wood, biomass) and modern alternative energy sources (solar, wind, water)."[442] Energy from conventional sources such as fossil fuels and nuclear are not included. "The scope is restricted to direct fuel

[441] The indicators are mainly derived and taken from P&G (2014): Excel sheet 2.

[442] P&G (2014): Excel sheet 2, incl. following sentence and quotation.

use and electricity generation onsite (as part of operations under the supplier's control) and the dedicated and certified purchase of renewable energy. It excludes the renewable energy that is part of the standard electricity grid at a given location." Besides the total amount of renewable energy, the energy intensity (2) might be tracked. While the share of renewable energy should be increased, the total energy consumption (also including renewable energy) should still be attempted to be decreased. Yet, in the beginning indicator (2) should rather increase as the share of renewable energy should increase, which will be measured with indicator (3), showing the ratio of the renewable energy consumption to total energy consumption (renewable and non-renewable).[443]

For recycling[444], the second optional category, it may be similar for DC suppliers as for renewable energy. Some suppliers may not have the technology, equipment or knowledge for recycling, and therefore the corresponding indicator cannot be applied. However, these indicators have their reason for existence, as it is widely expressed in literature, and therefore seen as relevant, and as it is also used by P&G, showing its applicability.[445] The absolute indicator (4) equals the total amount of all produced waste material that is not dumped in landfills or incinerators without any kind of energy recovery.[446] In other words, it comprises all material which has a useful life after being used, by getting reused or for incineration and composting/biogas to recover energy. It should increase in the beginning but overall the waste produced should decrease and therefore also the amount of waste recovered in the end should decrease. At the best all waste produced is recovered. The recycling rate (5)[447] shows how much of the total amount of waste material gets recycled, reused or recovered, and therefore showing the supplier's environmental performance improvement in these terms.

In order to assess the total GHG emissions of the company's suppliers it would be necessary to track also the indicator for Scope 3 GHG emissions (6).[448] However, as this involves a broad scope which may not be accessible to all, this is only an optional measure as well.[449] It includes all other indirect emissions that Scope 1 and 2 are not capturing. These emissions result from sources which are not directly under the control or in the possession of the firm but related to its corporate activities. Although these

[443] See P&G (2014): Excel sheet 2; Ahi (2015b), p. 2887: Non-renewable energy rate, comparable to renewable energy rate; Krajnc (2003), p. 283; GRI (2013), p. 90.

[444] The indicators are derived and taken from P&G (2014): Excel sheet 2.

[445] See P&G (2014): Excel sheet 2; Ahi (2015a), p. 366; Awasthi (2010), p. 374; Azadi (2015), p. 275; Erol (2011), p. 1093; Handfield (2002), p. 79f; Huang (2007), p. 521.

[446] See P&G (2014): Excel sheet 2, incl. following sentence.

[447] Krajnc (2003), p. 284.

[448] The indicators are mainly derived and taken from P&G (2014): Excel sheet 2.

[449] See WBCSD (2004), pp. 25, 29, incl. following sentences.

emissions are not directly generated by the supplier company, but through external emissions, they make up for a major part for some firms' GHG emissions.[450] The engagement in the reduction of these emissions is a claim for leadership and enhances innovation.[451] The supplier should be clear about which actions and emissions fall under the different scopes by referring to the respective consolidation approach.[452] The emissions resulting from the following activities are usually taken into Scope 3[453]: the "extraction and production of purchased materials and fuels", "electricity-related activities not included in scope 2", "transport-related activities", emissions from "leased assets, franchises and outsources activities", "use of sold products and services" and "waste disposal". The intensity indicator (7) should be again calculated for comparison reasons.

The fourth indicator, transportation fuel efficiency (8),[454] may also represent an important measure as the products from the suppliers have to be delivered to the purchasing firm. This transportation, from the supplier fabric to the customer, does also produce lots of CO_2 emissions that become part of the products bought by the firm. The indicator shows the efficiency as it shows how much grams of CO_2 are generated per ton-kilometre[455]. This number should be reduced over time, therefore improving sustainability. The mode of transport should be stated as well.

The last indicator, concerning fines and sanctions (9)[456], shows the "monetary value of [..] fines and total number of non-monetary sanctions for non-compliance with environmental laws and regulations."[457] Although suppliers should go beyond compliance, they first have to comply, and therefore this indicator gets tracked.

Apart from these quantitative measures the accounting department may assess additional information as they are already tracking the quantitative ones. Qualitative measures will supplement and complete the picture and evaluation of the suppliers. This assessment and tracking will be simplified with a specific tool, a report or scorecard, in this case a supplier scorecard, as P&G makes use of.[458] The qualitative measures may include, e.g. the assessment of sustainability ideas for the firm and

---

450 See GRI (2013), p. 112.

451 See GRI (2013), p. 112; WBCSD (2004), p. 75f.

452 See WBCSD (2004), p. 29.

453 See WBCSD (2004), p. 29, incl. all quotations and further details.

454 The indicator is derived from P&G (2014): Excel sheet 2, incl. following sentences; for further information of scope 3 transport-related activities see WBCSD (2004), pp. 29–31.

455 Tkm = Transported amount (t) * distance (km); see Goldmann (2001), p. 618.

456 The indicators are mainly derived and taken from P&G (2014): Excel sheet 2. See P&G (2014): Excel sheet 2; Handfield (2002), p. 79.

457 P&G (2014): Excel sheet 2.

458 See Handfield (2002), p. 84; Brown (1996), p. 121; for an example of a supplier scorecard see P&G (2014): Excel sheet 3.

supported initiatives by the suppliers which do also improve the firm's environmental footprint.[459] The share of suppliers with an environmental management system (EMS), which is also externally verified (such as ISO 14001) or the share of suppliers which have received environmental audits can also be measured with the help of corresponding information.[460] Another aspect is whether the supplier does have other certifications or follows relevant environmental standards.[461] The firm could also assess the suppliers' competences in the field of reverse logistics, so that the responsibility of disposal rests with the suppliers.[462]

Certainly, many more indicators have been detected and are available but the ones mentioned in this subsection have been considered as the most appropriate and relevant ones. In order to get the best results, the suppliers should be monitored on a regular basis, at least once a year, to receive relevant results, sufficient data for comparisons, to see improvements and achievements of environmental goals, and to support the goal to create an overall environmental value chain.[463] Site visits may also be reasonable from time to time, to check the reliability of the data provided.[464] It is reasonable to start this rather long-term project of implementation, with only a few important suppliers, following the Pareto principle or 80/20 rule, so going with the suppliers which make up for 80% of the firm's expenses but represent only 20% of the suppliers.[465] Of course, suppliers should not only be evaluated according to their environmental performance but also concerning overall criteria, including conventional indicators.[466] This can be achieved by following a "two step approach"[467], first evaluating the suppliers according to their environmental situation and then for standard criteria. For monitoring activities the environmental criteria may be assessed supplementary or combined, as it is preferred by the company.
Companies like Interface, Inc. and Hansgrohe, which are already active in the field of environmental indicators, still do not track their suppliers with environmental indicators, although Interface is making annual requests for respective environmental

---

459 See P&G (2014): Excel sheet 2.

460 Can be presented as percentages, and presents therefore a quantitative measure. See P&G (2014): Excel sheet 2; Singh (2011), p. 36; Rao (2014), p. 71; Rao (2008), pp. 45, 48, 55f; Rao (2009), p. 20; Goldmann (2001), p. 620, incl. following sentence.

461 See P&G (2014): Excel sheet 2; Awasthi (2010), p. 374; Handfield (2002), p. 79f; Charter (2001), p. 3.4.3.

462 See Handfield (2002), p. 79f; for reverse logistics see subsection 6.4.8.

463 See Handfield (2002), p. 84; Jasch (2000), p. 82f; Jasch (2009), p. 55; Akamp (2013), p. 56; Probst (2006), pp. 36–38; DIN (2013), pp. 31, 34; Colsman (2013), p. 67.

464 See Öztürk (2014), p. 133.

465 See Brown (1996), p. 121; here it may be more reasonable to go with the suppliers making up for 80% of material supply measured in mass units, not in monetary terms, as this is more responsible for the environmental impact than sales

466 See Awasthi (2010), p. 374.

467 Awasthi (2010), p. 374, incl. whole sentence.

information and Hansgrohe is planning for a supplier tracking.[468] This shows that there is still need for improvement.

6.4.2.4 Procurement and Inbound Logistics

While the previous subsection provided supplier specific indicators, this one will give possible indicators referring to procurement activities and inbound logistics excluding the direct evaluation of suppliers. Inbound logistics as the first part in a value chain comprises not only supplier evaluation, which has been discussed in the previous subsections, but also sourcing and purchasing of materials in general, the handling of purchased materials, as well as outsourcing, negotiation, delivery scheduling, and inventory management.[469] This phase is of high significance for minimizing the focal company's environmental impact (such as: waste reduction, avoidance of hazardous material and sourcing environmentally friendly materials[470]), as it cuts to the root of environmental issues, while leading to significant cost savings, better efficiency and quality output.[471] In this context, the procurement department shall minimize or prevent the buying of any hazardous and virgin materials, increase the amount of recyclable, recycled, reusable and reused materials, as well as increase bio-degradable and returnable packaging and reduce packaging overall.[472] It also concerns the shipment of products and materials, resulting i.a. in fuel consumption and emissions.[473] The transportation fuel efficiency has already been measured in the context of suppliers, which is more reasonable, as suppliers know their level of transport emissions best. Therefore, there is no need to calculate this indicator again. Except if the buyer company is responsible for the transportation of the goods to its own facilities.

Whereas it may be more reasonable to calculate the indicator of material consumption in the production department, as it gets actually consumed there, it is feasible to measure the purchase share of the different kinds of materials here.[474]

Material Shares

$$'Recycled'\ Material\ Share = \frac{Total\ Costs\ for\ 'Recycled'\ Materials\ (EUR)}{Total\ Costs\ for\ Material\ Procurement\ (EUR)}\ x\ 100 \uparrow\downarrow$$

---

468 See Hay (E-Mail); Hansgrohe SE (E-Mail).

469 See Porter (1985), p. 39f; Sarkis (1999), p. 6; Rao (2014), p. 58.

470 Environmentally friendly materials are "raw materials that have been produced using [...] recycled and/or reused materials." Rao (2008), p. 26f.

471 See Rao (2014), p. 58f; Rao (2005), p. 900; Rao (2008), p. 24f.

472 See Rao (2014), p. 59; Rao (2010), p. 2; Rao (2008), p. 26f.

473 See OECD (2011), p. 26.

474 The idea of the following indicator is derived from Rao (2010), p. 11; Rao (2014), pp. 59, 71; Rao (2008), pp. 26, 47.

The relative indicator should be applied to all different kinds of material shares, not only recycled material in relation to total materials purchased but also the proportion concerning recyclable, reused, reusable and (non-)/hazardous materials.[475] This indicator shows the percentage of the different kinds of materials in relation to the total material input measured in monetary terms. It is also important to define the period of measurement. Due to these indicators it is possible to see the extent to which the firm is already sourcing 'green' materials.[476] Percentage targets should be set for the respective period to be achieved. Whether the goal is to increase or decrease this indicator depends on the type of material share. While the recycled, reusable or recyclable share should be increased, the hazardous material share should be decreased.

6.4.3 Technology Development: Product Design

In order to green the whole value chain, the products (or services) should also be developed and designed in a manner to minimize the needed resources, reduce their negative environmental impact during their whole life-cycle, reduce waste in the final stage, and focusing on recycling, reusing and their reintegrability.[477] Eco-design enjoys a high level of recognition in research papers and literature.[478] "Environmental innovation consists of new and modified processes, equipment, prod-ucts, techniques and [..] systems that avoid or reduce harmful environmental impacts."[479] Environmental innovation can be intentional or unintentional, technical (involving new processes, technology and products), or organizational (encompassing structural change, such as making use of new tools, e.g. LCA). The product design process, "designing new or redesigning existing products" [480] is rather strict in Europe as not only regulatory but also environmental requirements have to be recognized. It should also address the design of packaging.[481]

As this value chain unit represents a support activity, it is strongly linked to primary activities, especially to operations but also in regards to marketing and sales. Thus, there is the possibility of overlaps, also in terms of indicators. For this reason, only specific divisional indicators shall be provided. The composition of products of

---

475 See Rao (2014), p. 71; Rao (2008), pp. 26, 47.

476 However, due to its calculation in monetary units, the results may be distorted, as green materials may cost more than unsustainable/conventional materials. Therefore, it is important to track the masses of materials used in operations as well. Yet, as resource depletion will increase, the costs of non-renewable materials may increase in future.

477 See Vijayvargy (2014), p. 27; Bask (2011), p. 18; Burschel (2001), pp. 269, 272; Singh (2011), p. 34; Yang (2011), p. 1182.

478 See Shen (2013), p. 175; Hashemi (2015), p. 180; Govindan (2013), p. 347; Hsu (2009), pp. 255, 257.

479 Arundel (2006), p. 325, incl. following sentence.

480 Askham (2012), p. 71, incl. whole sentence.

481 See Yang (2011), p. 1182f; Singh (2011), p. 34; Rao (2008), p. 70f.

different materials and the use of them refer to both production and R&D. Yet, in R&D it may be more interesting how many products are, for example, recyclable, and in production the mass may play a more important role.[482]

I. Product Design

$$\mathit{Percent\ of\ Products\ Designed\ for\ Disassembly, Reuse\ or\ Recycling} = \frac{\mathit{Number\ of\ Products\ with\ Specific\ Characteristic}}{\mathit{Total\ Number\ of\ Products}}\ x\ 100 \uparrow$$

The products should be divided into a list in terms of their nature, whether the "product or its parts are designed for disassembly, reuse or recycling."[483] Since not all products may be reusable to 100%, the products should be weighted concerning the degree of fulfilment of the parts of the product, meaning for instance, that if only 50% of a product is reusable, it should be weighted as 0.5 instead of 1.[484] These indicators, if compared over time, will show the progress in environmental development and the goal should be to increase this percentage.

The next indicators are highlighting the commitment of the firm in green design by showing the respective investment. Although the budget may be approved from general or financial management and not by the R&D department itself, it has the correct data of its investments and is responsible for them. That is why these indicators are assigned to this support activity and not to subsection 6.4.4.[485]

II. Investment in Environmental Innovation

Absolute Indicator:

$$(1)\ \mathit{Total\ R\&D\ Costs\ for\ Environmental\ Innovation\ (EUR)} \nearrow$$

Relative Indicator:

$$(2)\ \mathit{Percent\ of\ Investment\ in\ Environmental\ Innovation} = \frac{\mathit{Total\ R\&D\ Costs\ for\ Eco\text{-}Design\ (EUR)}}{\mathit{Total\ R\&D\ Costs\ (EUR)}}\ x\ 100 \uparrow$$

The total investment in eco-design (1) may concern expenditures for designing the products for disassembly, reuse or recycling but also the money invested in research for improving processes and methods, in order to reduce resource consumption or

482 See subsection 6.4.6.

483 Veleva (2001), p. 544, incl. whole sentence. The indicator is derived from Veleva (2001), pp. 526, 544; DIN (2013), p. 64; Krajnc (2003), p. 284.

484 See Veleva (2001), p. 544.

485 The indicators are derived from Krajnc (2003), p. 285; Epstein (2003), p. 29; Arundel (2006), pp. 327, 329.

improve pollution control[486]. It would be good to detect and calculate the intentional but also the unintentional innovations for the environment.[487] In this case all investment would be determined. The investment should be increased, if possible, especially in the beginning if not much is invested, but this may ease a little bit over time. Indicator (2) shows the share of R&D investment for green design against the total R&D costs, which should be increased. Intel Corporation is a good example of successful implementation of environmental criteria in this stage.[488]

6.4.4 Firm Infrastructure

The firm infrastructure according to Porter comprises i.a. general, financial and quality management (QM), which will be of greater importance in this subsection.[489] Referring to the firm's environmental impact, QM especially should not only take the responsibility for the general quality of products but also for reducing the environmental impact, also resulting in an increase in product quality in the end.[490] As QM is directly related to the ISO 9000 standards, it may be reasonable to entrust the QM with general certification.[491] In the context of environmental concerns it relates principally to the ISO 14000 series and some others which are related to environmental issues (e.g. RoHS, WEEE, EuP).[492]

I. Quality and General Management

Environmental Certification

$(1)\, Total\ Number\ of\ Certificates$ ↑

Labelling

$(2)\, Total\ Number\ of\ Eco\text{-}Labels$ ↑

Achievement of Environmental Objectives

$$(3) Degree\ of\ Target\ Attainment = \frac{Number\ of\ Achieved\ Environmental\ Targets}{Total\ Number\ of\ Set\ Environmental\ Targets} \times 100$$ ↑

486 See DIN (2013), p. 27; Epstein (2003), p. 29.

487 See Arundel (2006), p. 331.

488 See Rao (2008), p. 71.

489 See section 6.2.

490 See Singh (2011), p. 35.

491 See Viadiu (2006), p. 142f.

492 See Viadiu (2006), p. 142f; Yang (2011), p. 1184; the first indicator is derived from Epstein (2003), p. 29; Seichea (2009), p. 506; Yang (2011), pp. 1182, 1184.

Therefore, the total number and kinds of environmental certificates will be assessed (1). In case of a large organization it may be also possible to determine which and how many of its facilities are ISO 14000 certified in order to calculate the share of facilities with certification compared to the total number of facilities.[493] In this context it also makes sense to take the point of eco-labelling (2).[494] Although this topic might be seen in connection with marketing, it will not be in this paper.[495] Using eco-labels as a marketing tool implies that a firm is simply greenwashing and not taking it as a true organizational strategy. Thus, it was decided to designate this area to the responsibility of QM, which shall detect the number and kinds of eco-labels (2) used in the firm by possibly detecting which labels do have a positive environmental impact and which ones do not. It may also be reasonable for the firm, if it has defined specific environmental criteria for products, to calculate the share of products which align with the predefined environmental criteria, e.g. eco-labelled, or organic, in relation to the number of total products.[496] The prerequisite to calculate the last indicator (3) is that the firm has to set environmental targets it wants to reach, for example on a yearly basis.[497] With the help of this indicator the firm gets a better overview of whether it implements realistic targets, and whether it takes them seriously by achieving them.

II. General Management and Finance

Environmental Expenditures

$(1)\ Total\ Environmental\ Protection\ Investment\ (EUR)\ and\ by\ Type$

Compliance

$(2)\ Fines\ \&\ Sanctions = Total\ Amount\ of\ Monetary\ Fines\ (EUR)\ \&\ Total\ Number\ of\ Non\text{-}monetary\ Sanctions$ ↓

Savings

$(3)\ Total\ Amount\ of\ Savings\ through\ Environmental\ Practices$ ↑

[493] See Epstein (2003), p. 29; Seichea (2009), p. 506.

[494] The indicator is derived from Handfield (2002), p. 79f; Kinderyte (2010), p. 111; Rao (2005), p. 909. For further information concerning German eco-labels see Rubik (2001), pp. 281-288.

[495] See Dobin (2009), p. 42, incl. following sentence.

[496] See Jasch (2009), p. 61; Kinderyte (2010), p. 111; Goldmann (2001), p. 615.

[497] The indicator is taken from Goldmann (2001), p. 618; see DIN (2013), p. 55; Jasch (2002), p. 48.

The first indicator (1), concerning the environmental expenditures, shows the firms' commitment to environmental performance.[498] "Data on environmental performance measured against environmental mitigation and protection expenditures offers insights into how effectively the organization uses resources to improve performance."[499] Thus, showing how effective the firm's environmental investments have been. The investment data also supports the internal cost-benefit analyses. If the respective data will be assessed precisely over years it allows the firm to evaluate its investments in environmental improvement. As has been the case for the environmental costs concerning the supplier, the environmental costs (1) for the focal company should also be divided into environmental costs, concerning pollutant effects and remediation, as well as improvement and prevention costs.[500] The total environmental expenditures concerning the former two encompass at least the following expenditures: treatment and waste disposal costs, costs for treatment of emissions (e.g. filters, agents), water waste treatment costs, expenditures for emissions certificates, costs for equipment, maintenance, operating materials, services and personnel costs, insurance costs for environmental liability, clean-up costs and costs for energy consumption. The latter two expenditures which should be assessed are environmental management costs, at least including the following: investment in environmental training and education[501], external services for environmental management, R&D, additional expenditures to install cleaner technologies and to purchase green material, redesign products and costs for recycling. Costs concerning non-compliance fines are excluded. The required data may be received additional to the accounting through the procurement, HR and legal department.[502]

The expenditures concerning fines of non-compliance (2) are also an important indicator to see whether the firm does at least comply with environmental laws and regulations.[503] This is a prerequisite to be able to go beyond mere compliance. In addition, it addresses financial risks, and thus the firm's economic bottom line.[504] The monetary value of fines and the number of sanctions should at least comprise the non-compliance with the following: "international declarations, conventions, and treaties, as well as national, regional and local regulations" and "voluntary environmental

[498] The first indicator is taken from GRI (2013), p. 135; see Kinderyte (2010), p. 111; Krajnc (2003), p. 285; Rao (2009), p. 20; Campos (2015), p. 290.

[499] GRI (2013), p. 135, incl. following sentences.

[500] See subsection 6.4.2.3; GRI (2013), p. 135; Humphreys (2003), pp. 144f, 148, incl. following sentence which has adopted the key points stated by the sources.

[501] For the service industry in particular this may also include the promotion of public transport; see DIN (2013), p. 27.

[502] See GRI (2013), p. 135.

[503] The indicator is taken from GRI (2013), p. 131; see DIN (2013), pp. 23, 56; Veleva (2001), p. 525; Campos (2015), p. 290; Jasch (2002), pp. 45, 48.

[504] See GRI (2013), p. 131, incl. following sentences and two quotations.

agreements with regulating authorities". The accounting might find additional information from the legal department or the firm's EMS.

Although the last indicator (3) does not directly relate to improving the environmental situation, it indicates the positive effect of environmental performance on the firm's economic bottom line, and is an important motivation factor to continue with the environmental strategy and initiatives, especially for management.[505] It includes all savings achieved through environmental performance and improvement such as the reduction of resource consumption or the prevention of pollution or recycling and minimization of waste.

As these departments comprise all administrative duties, including the R&D, finance & accounting, legal, HR, marketing, sales and service department, it may also be reasonable to track the energy consumption (GJ), the water consumption (m^3), the material (mt) (especially the paper in mass or pieces) and the office waste per employee per year,[506] in order to receive the total amount of each. This is especially important to be able to calculate the energy, water and material consumption, and waste for the whole company for the KEPIs in section 6.5.

6.4.5 Human Resource Management

As the awareness and commitment of employees is critical with regard to environmental improvement, environmental education and training for employees are crucial activities.[507] The human resource management (HRM) is responsible for all activities and issues concerning personnel.[508] The following indicators should be tracked by environmental accounting.

505 The indicator is derived from DIN (2013), p. 57, incl. following sentence.

506 See Hopfenbeck (1993), p. 330; DIN (2013), p. 30.

507 See Yang (2011), p. 1184.

508 See subsection 6.4.6.

Hours of Environmental Training and Share

$$(1)\ Average\ Number\ of\ Hours\ of\ Environmental\ Training\ per\ Year =$$

$$\frac{Total\ Number\ of\ Hours\ of\ Environmental\ Training\ (hrs)}{Total\ Number\ of\ FTE} \uparrow$$

$$(2)\ Percent\ Trained\ Staff\ (Employees\ and\ Managers) =$$

$$\frac{Number\ of\ Trained\ Staff}{Total\ Number\ of\ Staff}\ x\ 100 \uparrow$$

Trainings

$$(3)\ Share\ of\ Environmental\ Trainings =$$

$$\frac{Total\ Number\ of\ Environmental\ Trainings}{Total\ Number\ of\ Trainings}\ x\ 100 \uparrow$$

Environmental Ideas & Suggestions

Absolute Indicator:

$$(4)\ Total\ Number\ of\ Environmental\ Suggestions\ by\ Employees \uparrow$$

Relative Indicators:

$$(5)\ Employee\ Suggestion\ Rate = \frac{Total\ Number\ of\ Environmental\ Suggestions}{Total\ Number\ of\ FTE} \uparrow$$

$$(6)\ Share\ of\ Environmental\ Suggestions =$$

$$\frac{Total\ Number\ of\ Environmental\ Suggestions}{Total\ Number\ of\ All\ Suggestions}\ x\ 100 \uparrow$$

The first two indicators refer to environmental training for employees.[509] While the first (1) shows the number of hours of training per employee, the second (2) presents the share of employees having attained environmental training, divided into employees and managers. This allows for a good overview of how many employees still need or have already received training and how much training has been received on average. If these indicators are tracked over time, the progress in environmental training and education can be well pursued.

[509] The first indicator is derived from Veleva (2001), pp. 526, 542f; Krajnc (2003), p. 285; Rao (2009), p. 20; the second indicator is derived from Seichea (2009), p. 506; DIN (2013), p. 56; Rao (2009), p. 20; Goldmann (2001), p. 619; Jasch (2000), p. 83.

In order to enhance these indicators it may also be required to increase the number of environmental trainings in relation to the total number of trainings, which the third indicator (3) is presenting.[510]

The last three indicators are all considering the possible suggestions and ideas of employees improving the environmental performance of the company.[511] These indicators present the total number of submitted suggestions (4), the number of environmental suggestions per employee (5) and the share of environmental suggestions against the total number of employee suggestions (6).[512] These indicators show the employees' environmental commitment, and whether the environmental strategy has been accepted, and how well it is truly ingrained in the firm's culture. Although these indicators should increase, it may be possible that the increase will decline over the years as with further progress it may become harder to suggest improvements. Yet, as greening the firm will probably be a long journey there might also be the continuous possibility for suggestions.

6.4.6 Operations[513]

Global energy and resource consumption, as well as the emission of GHG are to a large part affected by manufacturing practices.[514] "Green manufacturing is defined as production processes which use inputs with relatively low environmental impacts, which are highly efficient, and which generate little or no waste or pollution."[515] The design and production stage are strongly connected as products or components that produce waste or cannot be recycled or reused should be redesigned. Sustainable manufacturing can result in cost savings, better efficiency, lower safety expenditures, a better image, and eventually increase the economic bottom line.[516] This and the increase of environmentally friendly products, prevention of pollution and reduction of waste, can be achieved by making use of different production concepts such as closed loop manufacturing (RL), cleaner or lean production, reduction of (harmful) materials consumption, increase in recyclable and renewable materials, optimization of

---

510 The indicator is taken from Goldmann (2001), p. 619.

511 The fourth indicator is taken from DIN (2013), p. 56; Krajnc (2003), p. 285; the fifth indicator is taken from Veleva (2001), p. 541; the sixth indicator is taken from Goldmann (2001), p. 619.

512 See DIN (2013), p. 56; Veleva (2001), p. 541; Goldmann (2001), p. 619.

513 All stated indicators in this subsection will only be applied for production although they may also be applied for the whole company in general. However, this will be the case in section 6.5.

514 See Linke (2013), p. 556; Moneim (2013), p. 1. Operations comprise both the production of goods and services. Production can also be transferred to the service industry, e.g. the hotel industry. There, e.g. the kitchen and the different rooms make also use of resources. Yet, some indicators may be better or only applicable to the manufacturing industry.

515 Singh (2011), p. 34, incl. following sentence.

516 See Singh (2011), p. 34; OECD (2011), p. 6.

production processes and indicators.[517] The following indicators will support the measurement of the production activities' environmental impact.[518] Besides, it will help the business better understand green production, support decision-making concerning operations, allow for fulfilment of goals and facilitate comparisons.[519] The operations phase can be divided into inputs, operations themselves and outputs or products.[520] This subsection will divide the indicators into their effects on the different environmental media, comparable to subsection 6.4.2.3 that will simplify the aggregation to corporate environmental indicators for section 6.5.

The indicators and topics do not differ greatly to those for supplier evaluation as they represent the KEPIs for DC suppliers which do not differ a lot to the KEPIs of the buyer company. However, the indicators of production do only refer to the production stage and not to the whole company, like energy consumption, which might be highest in production but still should be applied to the whole company since others, for example administrative units do consume energy as well.[521]

---

[517] See Rao (2005), p. 902f; OECD (2011), p. 20; Moneim (2013), p. 1.

[518] See OECD (2011), p. 9.

[519] See Veleva (2001), p. 521f.

[520] See OECD (2011), p. 8; Appendix 9 and 10; for implementation of the indicators, which would exceed this paper, one can find information here: OECD (2011), p. 10ff.

[521] If it represents too much effort and costs to track the indicators for the main environmental media separately, they may also be tracked only for the whole firm as it is the case in section 6.5. Yet, it depends on the firm. Some may have the greatest effects in their operations and the indicator does not differ a lot to the KEPIs and may not be tracked in addition.

ENERGY[522]

As manufacturing consumes large amounts of energy, the energy consumption in production should be tracked.[523] The energy consumption may affect operational costs and face issues with energy supply and prices.[524] The environmental footprint depends strongly on the source of energy used. The indicators can help to detect savings potential and the main sources of energy.[525]

Absolute Indicator:

$$\begin{aligned}(1)\ &Total\ Direct\ Net\ Energy\ Consumption\ for\ Production\ (GJ)\\ &= Non\text{-}renewable\ Fuel\ Consumed\ (Purchased\ \&\ Generated)\\ &+ Renewable\ Fuel\ Consumed\\ \downarrow\quad &+ Electricity, Heating, Cooling\ and\ Steam\ Purchased\\ &+ Self\text{-}generated\ Electricity, Heating, Cooling\ and\ Steam\\ &- Energy\ Sold\end{aligned}$$

Relative Indicators:

$$(2)\ Source\ of\ Energy\ Rate = \frac{Consumption\ per\ Source\ of\ Energy\ (GJ)}{Total\ Energy\ Consumption\ (GJ)}\ x\ 100$$

$$(3)\ Renewable\ Energy\ Rate = \frac{Renewable\ Energy\ Consumption\ (GJ)}{Total\ Energy\ Consumption\ (GJ)}\ x\ 100\ \uparrow$$

$$(4)\ Energy\ Intensity = \frac{Total\ Direct\ Net\ Energy\ Consumption\ (GJ)}{Unit\ of\ Output}\ \downarrow$$

Indicator (1) adds up all sources of energy consumed in production.[526] The energy produced by the firm, not consumed in operations but sold to external parties will not

---

[522] The environmental issue and indicators have been proposed i.a. by OECD (2011), p. 27; OECD (URL2): O2. Energy intensity & O3. Renewable proportion of energy; Veleva (2001), pp. 525, 534; Krajnc (2003), p. 283; Jasch (2009), pp. 54, 59; Jasch (2002), p. 48; Ahi (2015a), p. 366; Campos (2015), p. 291; Müller (2011), p. 60; Rao (2009), p. 20; Kinderyte (2010), p. 111; Goldmann (2001), p. 602; DIN (2013), p. 61; Hopfenbeck (1993), pp. 341, 344; Verfaillie (2000), pp. 15, 20; Jasch (2000), p. 81; Moneim (2013), p. 5; GRI (2013), pp. 89–93; subsection 6.4.2.3. Consumption is a synonym to usage in the case of energy.

[523] See Moneim (2013), p. 5.

[524] See GRI (2013), p. 89, incl. following sentence.

[525] See Goldmann (2001), p. 602.

[526] The indicator is taken from GRI (2013), p. 90. See GRI (2013), p. 89f, incl. following sentences. All sources of energy should be considered, secondary (e.g. electricity, heat, fuels) and primary energy sources (e.g. hard coal, crude oil, natural gas, nuclear, waste, biomass, wind, hydro, tide). As secondary is derived from primary energy sources, it may be reasonable to account for both amounts consumed separately. Otherwise, it is possible to only account for primary energy use, also in case it is transferred to secondary energy, to avoid double counting, i.e. "When reporting self-generated energy consumption, the organization does not double-count fuel consumption. For example, if an organization generates electricity from coal and then consumes the generated electricity, the energy consumption is counted once under the fuel consumption." GRI (2013), p. 89. For this and further information see Øvergaard (2008); Arvidsson (2012); Riesner (2010), pp. 477–480; Hoffmann (2011), pp. 134-136, 147; GRI (2013), p. 89.

be included. The energy consumption should be again divided into electricity and fuel energy consumption and further divided into renewable and non-renewable energy, and all should be measured separately. Thus, all sources of energy and their amounts have to be identified. Non-renewable fuel sources comprise fuel "for combustion in boilers, furnaces, heaters, turbines, flares, incinerators, generators"[527]. Renewable fuels cover biofuels and biomass.[528] The energy consumption is measured in gigajoules. As the different energy carriers are measured in different metrics it is required to convert all energy data in the same metric.[529] Information sources may be utility bills and standards, and assumptions and methodologies have to be assigned, as well as the consistent application of the respective conversion factors.[530] Indicator (2) shows the share of each energy source in relation to the total energy consumption and will therefore allow for the identification of the main sources.[531] Although it could be part of indicator (2), indicator (3) presents in particular the percentage of energy from sustainable sources, as it may be of high importance for energy usage.[532] It should be increased in order to replace the conventional energy sources, such as fossil fuels, with renewable energy, such as biomass, biogas, tides, wind, solar or hydropower.[533] The last energy indicator (4) informs about the energy intensity, meaning the required energy for one unit of output.[534] Although it is possible to calculate product (per unit produced), service (per service) and sales intensities (per monetary unit of sales), the focus lies on the production output measured in the respective metric such as mt, m^3 or GJ, as it considers the production.[535] Yet, since it shall refer to the service industry as well, unit of output can also be translated into unit of service[536], meaning for example the working hours of employees for the respective reference period. The intensity indicator allows again for comparisons.

Interface, Inc. also measures its energy use at manufacturing sites, and with the help of eco-metrics they have determined that their energy use per unit of production declined by 40% from 1996 until 2014, and their renewable energy share has evolved to 45% of total energy use at manufacturing sites.[537]

---

527 GRI (2013), p. 89.

528 See GRI (2013), p. 90, incl. following sentence.

529 See Goldmann (2001), p. 602; for conversion see Appendix 11 and 12.

530 See Veleva (2001), p. 534; GRI (2013), p. 90.

531 The indicator is taken from Krajnc (2003), p. 283.

532 The indicator is taken from OECD (2011), p. 27; OECD (URL2): O3. Renewable proportion of energy.

533 See OECD (2011), p. 27; OECD (URL2): O3. Renewable proportion of energy; Veleva (2001), p. 534.

534 The indicator is derived from GRI (2013), p. 93; Krajnc (2003), p. 283; OECD (URL2): O2. Energy intensity.

535 See GRI (2013), p. 93. All intensity indicators concerning the production with unit of output in the denominator can comprise both 'per unit produced' and 'per unit of service'.

536 See Veleva (2001), p. 522.

537 See Hay (E-mail); Appendix 14.

WATER[538]

Large quantities of water are used for corporate operations, especially for agriculture and food production.[539] Therefore, this valuable natural asset should be tracked with special focus during the operations stage.

Absolute Indicators:

$$(1)\ Total\ Water\ Withdrawal\ for\ Production\ (m^3)\ \downarrow$$

$$(2)\ Total\ Water\ Discharge/Waste\ Water\ from\ Production\ (m^3)\ \downarrow$$

Relative Indicators:

$$(3)\ Volume\ Fraction\ of\ Water\ Type = \frac{Consumption\ per\ Water\ Type\ (m^3)}{Total\ Water\ Consumption\ (m^3)}\ x\ 100$$

$$(4)\ Reused\ and\ Reycled\ Water\ Fraction =$$

$$\frac{Total\ Volume\ of\ Water\ Reused\ \&\ Recycled\ (m^3)}{Total\ Volume\ of\ Water\ Consumed\ (m^3)}\ x\ 100\ \uparrow$$

$$(5)\ Water\ Intensity = \frac{Total\ Water\ Consumption\ (m^3)}{Unit\ of\ Output}\ \downarrow$$

$$(6)\ Waste\ Water\ Intensity = \frac{Total\ Water\ Discharge\ (m^3)}{Unit\ of\ Output}\ \downarrow$$

The first indicator (1)[540] measures the total water consumption/usage[541] in cubic metres (all volume metrics are conceivable) and comprises the total volume of water withdrawn from any kind of source, including surface water, ground water, rainwater directly collected by the company, waste water from other companies, or water from external water utilities that is used for production.[542] The consumption volumes of the different water sources should be measured separately in order to calculate indicator (3) which allows detection of the main sources.[543] If both absolute and relative indicators will be traced over time, the progress can be assessed well. The second indicator (2) implies the total amount of water discharged by the firm from production

---

538 The environmental issue and indicators have been proposed i.a. by OECD (2011), p. 27; OECD (URL2): O1. Water intensity; Veleva (2001), pp. 525, 533; Krajnc (2003), p. 283; Jasch (2009), pp. 54, 59; Jasch (2002), p. 48; Ahi (2015a), p. 366; Campos (2015), p. 291; Müller (2011), p. 61; Rao (2014), p. 71f; Rao (2008), p. 62; Rao (2009), p. 20; Kinderyte (2010), p. 111; Hopfenbeck (1993), p. 341; Goldmann (2001), pp. 607–609; Verfaillie (2000), pp. 15, 20; Jasch (2000), p. 81; DIN (2013), p. 61; Moneim (2013), p. 5; GRI (2013), pp. 97-99, 122.

539 See Sedghi (URL).

540 The indicator is derived from GRI (2013), p. 97; Krajnc (2003), p. 283.

541 As the total water intake is not always directly consumed in production but also used for e.g. cooling, heating or washing, the volume of usage and consumption might be tracked together. Consumption will include usage in this context. See OECD (URL2): O1. Water intensity.

542 See GRI (2013), p. 97; the total water intake does not refer to the total water intake for the whole company, such as water consumption of e.g. sanitary facilities, but only for production purposes. In the case of a service company, e.g. a hotel, the water consumption in the kitchen may be included for example.

543 The indicator is taken from Krajnc (2003), p. 283.

processes.[544] Waste water is often further processed for treatment concerning its potential load of nutrients and chemicals negatively affecting the environment. The total volume includes all planned and unplanned water discharged. If a water meter is not available, the indicator has to be estimated. This equals the subtraction of the approximate volume consumed in production from the volume withdrawn for production. Additionally, it is possible to assess the water quality of the effluents. Indicator (4) first needs to assess the total volume of water recycled and reused by the firm for production in order to put it in relation to the total volume of water consumed.[545] It includes all water that is reused (either treated or not treated before reuse) as well as grey water (the rainwater collected by the firm and household waste water).[546] This measure presents the firm's efficiency as it shows its capability of reducing the total water withdrawal and discharge. "Increased reuse and recycling may result in a reduction of water consumption, treatment, and disposal costs."[547] Water (5) and waste water intensity (6) measure the consumption of water and discharge per unit of output.[548] As it is the case for the other environmental issues, both should be measured, total volume and intensity.[549]
Interface Inc. was able to reduce its water intake per unit of production at its manufacturing sites by 87% within 18 years.[550] Hansgrohe SE also tracks its consumption of water as an internal indicator and makes use of the GRI indicators for sustainability reporting.[551]

---

544 The indicator is derived from GRI (2013), p. 122, incl. following sentences.

545 The indicator is derived from GRI (2013), p. 99.

546 See GRI (2013), p. 99, incl. following sentence.

547 GRI (2013), p. 99.

548 The indicators are derived from OECD (2011), p. 27; OECD (URL2): O1. Water intensity; Jasch (2009), pp. 54, 59; Krajnc (2003), p. 283f.

549 See OECD (URL2): O1. Water intensity.

550 See Hay (E-Mail); Appendix 14.

551 See Hansgrohe SE (E-Mail).

MATERIAL[552]

Although these indicators may be more important for the manufacturing industry than for the service industry, they should be provided, as they play an important role for production.[553] They determine the firm's material consumption and thereby its depletion of resources and the costs for operation.[554]

Absolute Indicators:

$$(1)\ Total\ Materials\ Used\ (mt\ or\ m^3)\ \downarrow$$

$$(2)\ Total\ Packaging\ Mass\ (mt)\ \downarrow$$

Relative Indicators:

$$(3)\ Mass\ Fraction\ of\ Material\ Type = \frac{Total\ Input\ Mass\ of\ Material\ Type\ (mt)}{Total\ Material\ Input\ Mass\ (mt)}\ x\ 100$$

$$(4)\ Material\ Intensity = \frac{Weight\ of\ Materials\ Consumed\ (mt)}{Unit\ of\ Output}\ \downarrow$$

Optional Product Indicators:

$$(5)\ Mass\ Fraction\ of\ Products\ from\ different\ Material\ Types = \frac{Mass\ of\ Products\ from\ specific\ Material\ Type\ (mt)}{Total\ Mass\ of\ Products\ (mt)}\ x\ 100$$

$$(6)\ Packaging\ Mass\ Fraction\ of\ Products = \frac{Packaging\ Mass\ (mt)}{Total\ Mass\ of\ Products\ (mt)}\ x\ 100\ \downarrow$$

$$(7)\ Mass\ Fraction\ of\ Reusable\ Packaging = \frac{Reusable\ Packaging\ Mass\ (mt)}{Total\ Packaging\ Mass\ (mt)}\ x\ 100\ \uparrow$$

In order to have a precise picture of the overall material consumption, all kinds of material consumed and used should be assessed, such as non-renewable and renewable raw materials (e.g. ores, minerals, wood), recycled and reused, auxiliary (required for manufacturing but not included in the end product, e.g. lubricants), operating, hazardous and toxic materials, materials for semi-products, components that are a part of the end product and materials for packaging.[555] These should all then be collected

552 The environmental issue and indicators have been proposed i.a. by OECD (2011), pp. 22, 34; OECD (URL2): I1, I3, P1-P4; Veleva (2001), pp. 525, 533; Jasch (2009), pp. 54, 59; Jasch (2002), p. 48; Campos (2015), p. 291; Müller (2011), p. 59; Rao (2005), p. 909; Rao (2014), p. 71; Rao (2008), pp. 47, 97, 104; Rao (2009), p. 20; Kinderyte (2010), p. 111; Hopfenbeck (1993), pp. 338, 340; Goldmann (2001), p. 610f; Verfaillie (2000), pp. 15, 20; Jasch (2000), p. 81; DIN (2013), p. 61; Krajnc (2003), p. 283; Moneim (2013), p. 5; GRI (2013), p. 86f.

553 Some service industries do also have high material consumption, such as the hotel industry, which needs for example a lot of food, which can also be seen as raw materials.

554 See GRI (2013), p. 86; Hoffmann (2011), p. 128.

555 The first indicator is derived from GRI (2013), p. 86; Jasch (2002), p. 48 adapted to the critique of Hoffmann (2011), pp. 128–131, which applies to the whole material section; the third and fourth indicator are derived from Krajnc (2003), p. 283; GRI (2013), p. 87; OECD (URL2): I1. Non-renewable materials

separately but added up to receive indicator (1), which is the basis for indicators (3) and (4). The metrics have to be converted in a single unit (here: mt)[556], if they are stated in volume and weight metrics.[557] Additional information to be collected may be whether the materials were bought from external parties or produced internally (e.g. captive production).[558] Concerning indicator (3), it might be of special interest to the firm to calculate the percentage of recycled or reused input materials, as this shows whether the firm is able to reuse them in their operations, and reducing virgin material consumption which greatly supports the environment.[559] However, all types of material can be regarded here separately, such as non- and renewable materials. Indicator (4) presents the "use of materials [...] as a proportion of production"[560], which should be reduced. Although some raw materials are abundant, such as iron ore, silver and copper, others are rather critical, such as rare earths, rhodium, platinum and manganese.[561] Water and fuels will be excluded for the material indicators. Besides the total mass of materials consumed, one can also relate it directly to the products generated in production. These indicators are called product indicators[562] and might be assessed additionally. The product indicators are directly related to the design stage.[563] Indicator (5) resembles indicator (3) but does not refer to the total mass of material input but to the total mass of products, showing the share of products which are from recyclable, recycled, reused and renewable materials or designed for disassembly, reuse or recycling.[564] To make the calculation more understandable it will be shown in the example of the recycled and reused content of products.[565] It is calculated as follows: The weight of one product unit times the proportion of recycled content times the units produced plus the weight of a product unit times the proportion of reused

intensity; OECD (2011), p. 22; see Müller (2011), p. 59f; Jasch (2002), p. 48; GRI (2013), p. 86, incl. following sentence.

556 As these indicators apply to a German firm again, it is also possible to state tons (t) instead of metric tons (mt). However, as this paper is in English it will further make use of mt.

557 See GRI (2013), p. 87.

558 See GRI (2013), p. 86.

559 See GRI (2013), p. 87; OECD (URL2): I3. Recycled/reused content; Hoffmann (2011), p. 130; the problem in this case is that the material reused in the reference period should not necessarily be related to the material used in the same reference period, as high fluctuations are possible. It would be better to relate it to the products' mass ending their product life in the reference period. Yet, this may be too difficult. For packaging, the reused packaging can be related to the packaging mass of the same reference period, as one can assume that the take back happens very prompt to the sales of the products; see Hoffmann (2011), p. 143.

560 OECD (2011), p. 22, incl. whole sentence. The indicator is taken from OECD (2011), p. 22; OECD (URL2): I1. Non-renewable materials intensity.

561 See OECD (URL2): I1. Non-renewable materials intensity, incl. following sentence.

562 See Goldmann (2001), p. 614f. If the mass is too much effort to be assessed, it is possible to account for the number of products and packaging; see Hoffmann (2011), p. 142f.

563 See OECD (URL2): P1. Recycled/reused content.

564 See Krajnc (2003), p. 284; the indicator is taken from Krajnc (2003), p. 284.

565 See OECD (2011), p. 34; the precise indicator is taken from OECD (URL2): P1. Recycled/reused content, incl. following sentences.

content times the units produced divided by the sum for each product (weight of a product unit times units produced) and this division will be multiplied by 100. Although indicator (5) looks similar to (3) it may vary a lot, as not all input materials which are used for production will directly become a part of the products. For example, some industries need more than 100 kilograms of raw materials to generate only one kilogram of product. Therefore, it may be reasonable to assess both indicators. The same applies for products with recyclable and renewable materials content.[566] As products are usually not sold without any packaging, the respective indicators (2), (6) and (7), may be also of great importance for firms.[567] Firstly, the total mass of packaging has to be assessed (2), including materials such as paper, cardboard or plastics[568]. From this, the packaging mass fraction (6) of the product can be derived, showing the share of packaging related to the total mass of products, which should be decreased. The last indicator (7) helps to see the percent of reusable packaging in relation to the total packaging mass, which should be increased. However, total packaging should be reduced in general.

In 2014, Interface achieved the use of 50% of recycled and biobased materials in carpet manufacturing.[569] Hansgrohe traces its reduction of hazardous materials.[570]

WASTE[571]

The situation in which a firm produces no waste and a closed-loop system is perfectly implemented, that is, all outputs of production become input again for operating processes, represents the perfect situation for the environment.[572] Waste affects the environment negatively, and beyond that, presents costs to the firm. The amount of waste implies the efficiency of the firm's production processes in comparison to its competitors.[573]

566 See OECD (URL2): P2. Recyclability & P3. Renewable materials content.

567 The indicators are taken from Krajnc (2003), p. 284, incl. following sentences.

568 See GRI (2013), p. 86.

569 See Hay (E-Mail); Appendix 14.

570 See Hansgrohe (E-Mail).

571 The environmental issue and indicators have been proposed i.a. by OECD (2011), p. 27; OECD (URL2): O5. Residuals intensity; Veleva (2001), pp. 525, 535; Krajnc (2003), p. 284; Jasch (2009), pp. 54, 58f; Jasch (2002), p. 48; Ahi (2015a), p. 366; Campos (2015), p. 291; Müller (2011), p. 60; Rao (2005), p. 909; Rao (2010), p. 11; Singh (2011), p. 36; Rao (2014), p. 71; Rao (2008), p. 47f; Rao (2009), pp. 20, 24; Kinderyte (2010), p. 111; Goldmann (2001), pp. 611–613; Verfaillie (2000), pp. 15, 21; Jasch (2000), p. 81; DIN (2013), p. 64; GRI (2013), p. 123.

572 See OECD (URL2): O5. Residuals intensity, incl. following sentence.

573 See Hoffmann (2011), p. 141.

Absolute Indicators:

(1) $Total\ Weight\ of\ Residuals\ (mt)$ – 2 Approaches: $\downarrow$

<u>a) Mass Balance Approach:</u>

$$Weight\ of\ all\ Inputs\ +$$
$$Weight\ of\ Fuel\ Consumed - Weight\ of\ all\ Products\ (mt)$$

<u>b) Waste Output Approach:</u>

$$Weight\ of\ Releases\ from\ Production\ Processes\ to\ Air\ +$$
$$Weight\ of\ Releases\ to\ Surface\ Water\ +\ Weight\ of\ Releases\ to\ Land\ +$$
$$Weight\ of\ Releases\ from\ Landfills\ +\ Weight\ of\ Transfers\ to\ Disposal\ +$$
$$Weight\ of\ Transfers\ for\ Treatment + Weight\ of\ Transfers\ to\ Recycling\ +$$
$$Weight\ of\ Transfers\ for\ Energy\ Recovery\ +$$
$$Weight\ of\ Transfers\ to\ Sewage\ +\ Weight\ of\ additional\ GHGs\ Produced\ +$$
$$Carbon\ Content\ of\ Direct\ Energy\ Use\ (mt)$$

(2) $Total\ Weight\ of\ Hazardous\ Waste\ and\ Non\text{-}Hazardous\ Waste\ (mt)$

(3) $Total\ Weight\ of\ Waste\ for\ Recycling\ and\ Disposal\ (mt)$

Relative Indicators:

$$(4)\ Waste\ Intensity = \frac{Total\ Weight\ of\ Waste\ (mt)}{Unit\ of\ Output}\ \downarrow$$

$$(5)\ Recycling\ Rate = \frac{Total\ Weight\ of\ Recycled\ Waste\ (mt)}{Total\ Weight\ of\ Waste\ (mt)}\ x\ 100\ \uparrow$$

In order to gain an overview of how the weight of waste can be reduced, the total weight should be calculated with the help of indicator (1), (2) and (3).[574] The first indicator comprises all possible releases, also those to the air and additional GHG which will be assessed separately in the area 'emissions'.[575] Yet, as this indicator is very comprehensive and some parts may be very difficult or insignificant for some to track it can be adapted accordingly. Concerning the two approaches, both should be applied as it is possible that the mass balance approach detects sources of residuals which have been forgotten with the waste output approach. The releases do not only comprise chemicals and metals but all materials perceived important such as organics,

574 The first indicator is taken from OECD (URL2): O5. Residuals intensity; the second and third indicator are derived from GRI (2013), p. 123; P&G (2014): Excel sheet 2.

575 See OECD (URL2): O5. Residuals intensity, incl. following sentences.

plastics or paper. In order to be consistent the indicator should be measured in metric tons. This shows which residuals are most significant in quantity terms and should be focused on. Indicators (1) and (2) will be calculated separately as it is important to detect the hazardous waste, which is "defined by national legislation at the point of generation"[576]. Non-hazardous waste includes all other forms of waste, liquid or solid except for waste water which has been accounted for in the area of 'water'. If possible, these indicators should also be tracked according to their disposal method (3), such as reuse, recycling, composting, recovery, incineration, deep well injection, landfill, on-site storage and other disposal methods[577]. These indicators are required in order to calculate the respective waste intensity (4)[578] and the total, allowing for important comparisons. Especially hazardous waste intensity gives information about the weight of dangerous waste per unit of output. Thus, not only waste in total but especially hazardous waste should be reduced by the firm. In order to reduce the absolute weight of waste coming from production processes the firm should try to recycle in order to minimize the produced waste and the use of virgin material. The share of recycled or reused waste will be determined through indicator (5)[579], showing the firms' competence in recycling. The denominator should include the total weight of all kinds of waste.

Interface makes use of the indicator 'waste to landfill' from manufacturing sites per unit of production which it was able to reduce by 91% since 1996, and since 1995 it diverted 140,000 tons of carpet from landfills.[580]

EMISSIONS[581]

Emissions are the last core environmental issue which should be tracked. Although emissions are already included in the total residuals indicator, air pollutants should be traced separately since they affect the environment greatly, including the climate, acidification, air quality (smog), agriculture, health and habitats.[582]

---

576 GRI (2013), p. 123, incl. whole sentence and following sentences.

577 List has been adopted from GRI (2013), p. 123.

578 The indicator is derived from Krajnc (2003), p. 284.

579 The indicator is derived from Krajnc (2003), p. 284.

580 See Hay (E-Mail); Appendix 14.

581 The environmental issue and indicators have been proposed i.a. by OECD (2011), p. 27; OECD (URL2): O6. Air releases intensity; Krajnc (2003), p. 284; Jasch (2009), pp. 54, 59; Ahi (2015a), p. 366; Rao (2010), p. 11; Jasch (2002), p. 48; Rao (2014), p. 71; Rao (2008), p. 48; Rao (2009), p. 20; Kinderyte (2010), p. 111; Goldmann (2001), p. 605f; Verfaillie (2000), pp. 15, 21; Jasch (2000), p. 81; Moneim (2013), p. 5; DIN (2013), p. 65; GRI (2013), pp. 107–120; Veleva (2001), pp. 525, 535f; Singh (2011), p. 36; Goldmann (2001), p. 605f.

582 See GRI (2013), p. 119; OECD (URL2): O6. Air releases intensity.

Absolute Indicator:

$$(1)\ Total\ Amount\ of\ Emissions\ (mt\ of\ CO2e) \downarrow$$

Relative Indicator:

$$(2)\ Emissions\ Intensity = \frac{Total\ Amount\ of\ Emissions\ (mt\ of\ CO2e)}{Unit\ of\ Output} \downarrow$$

For this environmental area it is necessary to track the production's total amount of emissions (1)[583], predominately the GHG emissions[584] in metric tons of CO_2 equivalents but it may also be reasonable to track ozone-depleting substances (ODS) in CFC-11 equivalent or other significant air emissions, such as NO_x, SO_x, persistent organic pollutants (POP), volatile organic compounds (VOC), hazardous air pollutants (HAP) or particulate matter (PM), depending on the specific situation.[585] As the emissions shall only be tracked for production activities, the different scopes for GHG emissions do only apply if they are directly related to the production stage.[586] Yet, as was already necessary for the emissions indicators in subsection 6.4.2.3, it is important to apply a consistent consolidation approach for the emissions and a base year.[587] The emissions intensity (2) can then be applied to the different types of air emissions or only to the total of GHG emissions, for instance, if they are the only emissions significant to the company.[588] The intensities present the emissions per unit of output, showing the normalized impact during the period tracked.[589]

Interface, Inc. has tracked its GHG emissions per unit of production since 1996 and was able to decrease them by 73% until 2014.[590] It is also measuring its product carbon footprint of carpet, which is down by 22% on average since 2008. Hansgrohe SE measures also CO_2 emissions for internal purposes.[591]

583 The indicator is derived from GRI (2013), p. 107; Goldmann (2001), p. 605f.

584 Comprising the same emissions compared to subsection 6.4.2.3 or GRI (2013), p. 107.

585 See GRI (2013), pp. 107–120, also for further information; Rao (2014), p. 71.

586 See GRI (2013), p. 107.

587 See GRI (2013), p. 108.

588 The indicator is derived from GRI (2013), p. 115; P&G (2014): Excel sheet 2; OECD (2011), p. 27; OECD (URL2): O6. Air releases intensity.

589 See GRI (2013), p. 115; OECD (URL2): O6. Air releases intensity.

590 See Hay (E-Mail); Appendix 14, incl. following sentence.

591 See Hansgrohe SE (E-Mail).

Optional: **BIODIVERSITY**[592]**: Land**[593]

Although this aspect is significantly affected by unsustainable behaviour, it is rather difficult to account for it separately, as most of the other environmental areas have negative effects on the biodiversity as well. Yet, this section will refer to the affected land of production sites. The indicators are declared as optional[594], as they cannot be directly applied to all kinds of firms, like the service industry, concerning manufacturing sites but they may be applied for administrative offices or other facilities in this case. Besides, only bigger firms with several facilities or the ones who plan for expansion have a real impact on land.

Absolute Indicator:

(1) $Total\ Surface\ of\ Facilities\ in\ or\ close\ to\ Protected\ and\ High\ Biodiversity\ Areas\ (m^2)$ ↓

Relative Indicator:

(2) $Natural\ Cover = \frac{Natural\ Cover\ Area\ (m^2)}{Total\ Land\ Area\ (m^2)}\ x\ 100$ ↑

The first indicator (1) refers to the total area of the company's facilities in square metres, covering areas which are protected or of high biodiversity.[595] The sites should be owned, leased or managed by the firm. It may give an incentive to future avoidance or to try to minimize the sites' area or number, in order to improve the impact on the biodiversity and avoid risks for the corporate image. The second indicator (2) implies almost the opposite.[596] It shows the share of the firm's land area that presents natural cover area.[597] Natural cover comprises "undisturbed land, as well as landscaped green cover, including 'green roofs'. In the case of undisturbed land, it could include natural forest, grassland, scrub or wetland." It expresses the firm's intensity of land used. Although a high share of natural land cover is rather improbable, the firm may try to use less area for its facilities. As the indicator assesses the situation of the land that is already covered, the firm is challenged to find ways to improve the current situation.

All the presented indicators for production shall enable the firm to green its complete production process and operations in general, as for example the Rasa Sayang Resort Hotel was able to with the help of an EMS.[598]

592 Defined as the number of species in a specific geographical area: Begon (2009), p. 544.

593 The environmental issue and indicators have been proposed i.a. by OECD (2011), p. 27; OECD (URL2): O8. Proportion of natural land; Moneim (2013), p. 5; GRI (2013), p. 100f.

594 It is also optional as the possible improvements may not be as large and simple as for other areas.

595 The indicator is taken from GRI (2013), p. 101; see GRI (2013), p. 101, incl. following sentences.

596 The indicator is taken from OECD (2011), p. 27; OECD (URL2): O8. Proportion of natural land.

597 See OECD (URL2): O8. Proportion of natural land, incl. following quotation and sentences.

598 See Rao (2008), pp. 87–95.

6.4.7 Outbound Logistics

Outbound logistics concerns activities such as the delivery and transport of products and services as well as packaging and waste management, like disposal.[599] Contrary to pollution prevention concerning the inbound phase, this phase tries to minimize the negative environmental impact resulting from earlier phases as well as transportation effects. Logistics decisions are often compromised by trade-offs between "direct shipping or hub-and-spoke, central warehouse or distributed network, intermodal or single mode, and third party services or private fleet."[600] Improving the environmental impact may result in less shipments and handling, shorter distances and better occupancy but on the other hand it may have negative impacts on time, flexibility, quality and costs. Packaging has a high impact on products' transportation, as it may minimize materials usage and increase the utilization of trucks, as well as reducing handling.[601] In order to develop an environmentally friendly transportation system, different issues have to be considered, such as sources for fuel consumption, infrastructure, the type of vehicle and the mode of transport (where railway should be preferred to the road as it is most efficient).[602] Transportation especially can have negative impacts on the environment through energy consumption, emissions, effluents and waste when transporting goods, products, materials, as well as employees.[603] This is why the following indicators will refer to transportation only.

[599] See Rao (2014), p. 59, incl. following sentence.

[600] Sarkis (1999), p. 16, incl. whole sentence and following sentence.

[601] See Singh (2011), p. 35.

[602] See Rao (2005), p. 905; Rao (2008), p. 106; Singh (2011), p. 35.

[603] See GRI (2013), p. 133; these negative effects can also be caused by inbound logistics. The indicator would only be created for the buyer firm if the transportation from the supplier to the buyer is led by the buyer. Here, it will be assumed that the buyer is only directly responsible for the transport to its own customers. See subsection 6.4.2.3: 'Transportation Fuel Efficiency'.

I. Transportation

Traffic Volume & Mode of Transport

Absolute Indicator:

$$(1)\ Traffic\ Volume\ per\ Transport\ Mode\ (km\ or\ tkm)\ \downarrow$$

Relative Indicators:

$$(2)\ Share\ of\ Transport\ Mode = \frac{Volume\ of\ Transport\ Mode\ (km\ or\ tkm)}{Total\ Traffic\ Volume\ (km\ or\ tkm)}\ x\ 100$$

$$(3)\ Transport\ Intensity = \frac{Total\ Transport\ Volume\ (km\ or\ tkm)}{Product\ Output\ (mt\ or\ units)}\ \downarrow$$

Energy

$$(4)\ Total\ Energy\ Consumption\ (GJ)\ of\ Vehicles\ \downarrow$$

$$(5)\ Share\ of\ Environmentally\ Friendly\ Vehicles =$$

$$\frac{Traffic\ Volume\ of\ Environmentally\ Friendly\ Vehicles\ (tkm)}{Total\ Traffic\ Volume\ (tkm)}\ x\ 100\ \uparrow$$

Emissions

$$(6)\ Total\ Emissions\ (mt\ of\ CO2e)\ \downarrow$$

The first three indicators give an insight into the firm's transportation infrastructure and intensity.[604] They show the total traffic volume for each transport mode (1) and the percentage of each transport mode on the total transportation respectively (2), as well as the transport intensity (3).[605] The different modes of transport may be trucks, trains, ships, planes or pipelines.[606] The traffic volume can be calculated in kilometres (distance travelled) or ton-kilometres.[607] The first indicator should be reduced for each transport mode, while the shares of transport modes may indicate which should be reduced if they present the ones with the highest environmental impact. The intensity indicator should be reduced, showing the distance which accrues for the transported goods.[608] It can be calculated for each single product and summed up to the total transport intensity, showing the distance for each product unit and for the total annual

---

604 The first two indicators are taken from Goldmann (2001), p. 618; see Björklund (2012), p. 36; Jasch (2009), p. 61; the third indicator is taken from Goldmann (2001), p. 618; see Flämig (2001), pp. 363, 365.

605 See Flämig (2001), pp. 363, 365; Goldmann (2001), p. 618; Björklund (2012), p. 36; Jasch (2009), p. 61.

606 See Goldmann (2001), p. 618; Jasch (2009), p. 61; Singh (2011), p. 35.

607 See Goldmann (2001), p. 618.

608 See Flämig (2001), pp. 365, 368, incl. following sentence.

production. The indicators including ton-km may also be applied for service firms with mainly passenger transport, where tkm will be replaced by person-kilometre.[609]

The first energy indicator (4) shows the total consumption of energy for vehicles, "such as oil, kerosene, fuel or electricity"[610]. The respective sources of energy should be converted into a unified metric, i.e. gigajoules.[611] The goal is to reduce this consumption over the measurement periods. Indicator (5) shows the share of vehicles that are environmentally friendly, meaning the vehicles that make use of renewable energy, electricity or natural gas.[612] This share is calculated by the traffic volume of vehicles consuming environmentally friendly sources divided by the total traffic volume times 100. This indicator should rise over the years.

The emissions indicator (6) will calculate the air emissions caused by the firm's transportation activities, not only concerning goods but also person transportation, which is in particular important for the service industry.[613] It may include all the air emissions which can be assessed, such as the GHG emissions.[614]

6.4.8 Marketing, Sales and Service

Although environmental management might be associated with marketing in terms of promoting the firm's green image[615], this is not the aim of this paper. Green marketing is especially focused on assessing and reacting to the behaviour of customers, in this case, green consumer, and educating them in terms of benefits and use of green products.[616] Customer service will be connected to reverse logistics in this case, considering the end-of life management of products concerning their reuse and return.[617]

---

609 See Goldmann (2001), p. 618.

610 GRI (2013), p. 133, incl. whole sentence; the fourth indicator is derived from GRI (2013), p. 133; Flämig (2001), p. 363; DIN (2013), p. 27; the fifth indicator is derived from Rao (2014), p. 72; Rao (2008), pp. 106–108.

611 It should be the same for all energy indicators. For conversion tables see Appendix 11 and 12.

612 See Rao (2014), p. 72; Rao (2008), pp. 106–108.

613 The indicator is derived from GRI (2013), p. 133; Flämig (2001), p. 363; DIN (2013), p. 27; WBCSD (2004), p. 27; Björklund (2012), p. 36.

614 See GRI (2013), p. 133.

615 See Ahi (2015a), p. 371; Tuzkaya (2009), p. 486; Awasthi (2010), p. 371.

616 See Rao (2008), p. 97.

617 See Bask (2011), p. 22.

I. Marketing & Sales

Press

$$(1)\ Total\ Number\ of\ Press\ Reports\ on\ Firm's\ Environmental\ Performance$$

$$(2)\ Share\ of\ Positive\ Reports = \frac{Total\ Number\ of\ Positive\ Reports\ per\ Year}{Total\ Number\ of\ Reports\ per\ Year}\ x\ 100 \uparrow$$

Customer

$$(3)\ Share\ of\ Green\ Customers = \frac{Total\ of\ Green\ Customers\ (EUR)}{Total\ Customers\ (EUR)}\ x\ 100 \uparrow$$

Marketing shall not serve as a superficial tool to greenwash and promote a firm's green image if there is none, but it shall use its expertise in market analyses. The press indicator (1) refers to the firm's environmental appearance in public.[618] The total number per year of positive and negative papers concerning the company's environmental performance should be determined in order to not only gain information about the internal environmental performance but also about the external effect on people, as well as whether the positive environmental impressions outweigh any negative ones with the help of the relative indicator (2). The indicator referring to the green customer ratio (3) shows the percentage of customers preferring eco-products measured in sales unit.[619] This indicator can also be renamed to the revenue fraction of eco-products, therefore calculating the total revenues from eco-products and then dividing those by the total revenues per year, for instance.[620] For these indicators, eco-products have to be defined clearly, what they comprise, e.g. organic and eco-labelled products[621], or being produced in an environmentally friendly way. This shows the popularity and current importance of green products in relation to conventional products, and it may also indicate whether the firm has to enhance promotion and/or better educate the customers about green products. The hotel Mandarin Oriental in Manila, for instance, has undertaken a survey among its customers in order to detect their demand for environmentally friendly products.[622]

---

618 The first two indicators are derived from DIN (2013), p. 57; Campos (2015), p. 291; Krajnc (2003), p. 285.

619 The indicator is derived from Hashemi (2015), p. 180; Krajnc (2003), p. 284; Awasthi (2010), p. 374; Shen (2013), p. 175.

620 See Goldmann (2001), p. 616; Krajnc (2003), p. 284; Jasch (2002), p. 48.

621 See Jasch (2002), p. 48; Kinderyte (2010), p. 111.

622 See Rao (2008), p. 102.

Reverse Logistics

Reverse logistics (RL) has gained in high significance over the last years, and many companies show progress in this field, such as Xerox in particular.[623] In the case of RL the supplier of products to the end consumer should be responsible for the return, collection and disposal of products and packaging from the customers.[624] It comprises therefore the return to "the 'forward logistics' chain"[625], in order to remanufacture or recycle materials and prevent waste to landfills as well as reducing the demand for virgin materials and resources, while being profitable at the same time.[626] In the EU there are some regulations which obligate firms to reuse and collect their own products from customers.[627] Although it is directly connected to the operations phase it will be seen as a customer service in this case.

II. Service

Reverse Logistics

(1) $\mathit{Share\ of\ Products\ with\ Take\text{-}back\ Policies} =$

$$\frac{\mathit{Number\ of\ Products\ with\ Take{-}back\ Policies}}{\mathit{Total\ Number\ of\ Products}}\ x\ 100 \uparrow$$

(2) $\mathit{Share\ of\ Reclaimed\ Products\ and\ Packaging\ Materials} =$

$$\frac{\mathit{Products\ and\ Packaging\ Reclaimed\ in\ the\ Reporting\ Period}}{\mathit{Products\ Sold\ in\ the\ Reporting\ Period}}\ x\ 100 \uparrow$$

The first indicator (1) presents the share of products which are theoretically for return, while the second indicator (2) shows the share of products which have been actually reclaimed.[628] For the first indicator the company should create a list showing all products with return policies and divide it then by the total number of products.[629] Although actually taking back the products might still be a big challenge for firms, it is worth it, as it increases the efficiency in terms of material and resource usage, being a source of differentiation, and reducing disposal costs and waste.[630] The second indicator (2) provides information concerning the firm's competence in taking back its products and materials, and transforming them into new materials for usage.[631] It counts the amount of products including their packaging materials which are

623 See Shaik (2012), p. 23; Singh (2011), p. 35; for the Xerox example and others see Rao (2008), pp. 120–128.

624 See Yang (2011), p. 1183; Rogers (1999), p. 2; Rao (2014), p. 59; Singh (2011), p. 35.

625 Sarkis (1999), p. 17.

626 See Guide (2001), p. 3; Singh (2011), p. 35f.

627 See Guide (2001), p. 3.

628 The first indicator is taken from Veleva (2001), p. 545; the second indicator is taken from GRI (2013), p.129; see Rao (2014), p. 72; Rao (2008), p. 117; Kinderyte (2010), p. 111.

629 See Veleva (2001), p. 545.

630 See Rao (2008), pp. 117–119; GRI (2013), p. 129.

631 See GRI (2013), p. 129, incl. following sentences.

reclaimed, meaning recycled and reused within the determined reporting period. Product rejections and recalls will be excluded and the recycling and reuse of packing should be calculated as an extra. The metric for the nominator and denominator is the amount (number) of products or sales in euro in this case, as the production output will be measured in the operations stage, which can also be directly linked to reverse logistics.

6.5 Corporate Environmental Indicators (KEIs)

The proposed environmental indicators may be beneficial for the support and tracking of divisional managers but all together may be still too much and too detailed for senior executives. Although the indicators have already been minimized to a relatively small number, as already mentioned, top level managers do not have the time to regard all non-crucial indicators. Therefore, there is a need for indicators that reflect the corporate environmental situation, also stated as KEPIs[632]. KEPIs should be defined individually according to the firm's environmental strategy.[633] As this paper tries to find key environmental indicators (KEIs)[634] across all sectors and companies, the following KEIs are based on the findings of the previous sections, and try to encompass all important environmental issues and value chain units in order to actually green the whole business. Yet, depending on the specific situation of the company, they might be adapted respectively.[635] As the indicators are intended to inform not only accounting but especially top-level management, it should not only provide EPIs but also EMIs, as managers may be divorced from these issues in physical terms, such as tons of waste, but rather the waste disposal costs or savings are more relevant to them.[636] Nevertheless, managers should also try to gain a general understanding of environmental issues in physical terms as this might be inevitable if applying environmental indicators. The following thirteen indicators are a suggestion for possible corporate environmental indicators. In general, it may be problematic for some firms to collect all required information. Thus, the value of each indicator should be assessed, and whether the firm is able to implement the respective indicator.

The first six indicators are comparable to the indicators provided for supplier evaluation as those have already demonstrated core indicators as well as the operations indicators, since they encompass all five environmental medias as well. The first six indicators represent the total input and output indicators for the whole company in the

632 See Doublet (2014), p. 1; Singhal (2004); Pérez-Calderón (2012), p. 1008.

633 See PwC (2006), pp. 5, 8; Brown (1996), p. 5; Probst (2006), p. 20f.

634 Here referred to as KEIs not KEPIs, as it includes not only EPIs but also EMIs.

635 Therefore, the following indicators will not be defined in detail, as they have already been defined more precisely or have to be adapted by the company accordingly.

636 See Jasch (2009), p. 58.

respective physical metrics. However, as it has already been mentioned the corresponding costs might be assessed as well, as this is often of major concern for management. In addition, the intensities for the six indicators might be calculated, as this will allow for better comparisons at this level. Besides, it may be of special interest for the management to see the reduction rates of the first six indicators. The reduction rate can be calculated by dividing, for example, the total net energy consumption of the current period by the total net energy consumption of the last period multiplied by 100.[637] This shows the general progress but neglects the fact that an increase or decrease may also be connected to a decrease or an increase in production output. Yet, the following indicators represent the key indicators in this context.[638]

$$(1)\ Energy\ Consumption\ (GJ)$$

Indicator (1) refers to the total energy consumption measured in gigajoules.[639] This includes the complete corporate energy usage, including energy for transportation but excluding the energy consumption of suppliers[640], including renewable and non-renewable, fuel and electricity subtracted by the energy sold.[641] The indicator should be reduced over the reference periods. If it is of special interest for the firm, the share of renewable energy consumption may be calculated as well.

$$(2\&3)\ Water\ Withdrawal\ \&\ Water\ Discharge\ (m^3)$$

Indicators (2&3) provide information for the second environmental media, water and waste water. Not only will the water intake[642] be assessed but also the water discharge (waste water). The types of water included have already been defined in subsections 6.4.2.3 and 6.4.6. In the case of water it may be also of interest to calculate the share of water reused or recycled in relation to the total volume of water consumed.[643]

[637] Calculation is derived from the 'Waste Reduction Rate' from ICAEW (URL).

[638] The selection of the following indicators is based on the selected indicators for the whole value chain, including the suppliers indicators, and supported by the following sources ICAEW (URL); Verfaillie (2000), pp. 15, 20f; Rao (2009), p. 17; WBCSD (2004), pp. 24–29; Schwarz (2002), p. 58; Gallego-Álvarez (2012), p. 255; Peña (2013), p. 5; GRI (2013); McIntyre (1998), pp. 153–155.

[639] For the formula see 6.4.6 but for this indicator all energy consumed and used will be included, not only the one in production but also for running the whole company, including i.a. the offices.

[640] This will be measured separately with the help of indicator (12).

[641] See GRI (2013), p. 89.

[642] Like in subsection 6.4.6, water withdrawal means that not only water consumption will be measured but also water usage which will end up in waste water and which is not included in a product, such as heating or cooling water, but also water for sanitation, cleaning or cooking.

[643] See GRI (2013), p. 99.

$$(4)\ Material\ Consumption\ (mt)$$

The material input (4) for the whole firm comprises all kinds of materials as already mentioned in subsection 6.4.6, but also including materials like paper for the offices. Here it may also make sense to calculate the share of recycled and renewable material in relation to the total material input.[644] As the material indicator is also closely connected to product indicators, it may also be meaningful to calculate the product recycling rate.[645] It presents the percentage of the total amount of products recycled divided by the total amount of products sold, multiplied by 100. The indicators concerning recycling or renewable resources are important as they show the progress towards a transformation in an environmentally sustainable business, and therefore demonstrate the acceptance of profound change and not only reduction efforts.

$$(5)\ Waste\ Production\ (mt)$$

The total amount of all kinds of waste (5) coming from all divisions should be calculated by applying the two different approaches[646], and may be further divided into hazardous and non-hazardous waste, and waste for disposal and recycling.[647] The waste recycling rate may be calculated in addition.[648] The costs for waste disposal (7) play an important role for the management as they can see how much they invest in waste, instead of trying to recycle or minimize it.

$$(6)\ GHG\ Emissions\ (mt\ of\ CO2e)$$

The total GHG emissions or the carbon footprint (6), comprise all scopes of GHG emissions.[649] They are measured separately, in direct (Scope 1) and indirect (Scope 2) emissions, while Scope 3 is optional, as it may not allow for comparisons, and as it is not that easy to collect the required data, however it may be meaningful as well.[650] It includes also one dimension which has not been addressed so far, the customer stage, by also accounting for the emissions coming from the sold product and service usage.[651]

644 See GRI (2013), p. 87.

645 See ICAEW (URL), incl. following sentence.

646 See subsection 6.4.6.

647 See OECD (URL2): O5. Residuals intensity; GRI (2013), p. 123.

648 See ICAEW (URL).

649 See subsection 6.4.2.3; WBCSD (2004), pp. 25–31.

650 See WBCSD (2004), pp. 25–31.

651 See WBCSD (2004), p. 29.

$$(7)\ Environmental\ Costs\ (EUR)\ \&\ Fines\ (EUR)$$

As managers are especially concerned with costs, all investments in environmental improvement but also the costs for energy consumption, water intake, waste water treatment, material, waste disposal and emissions treatment, indicator (7) should be provided, as well as the costs for fines, as these show the necessity for environmental improvement.

$$(8)\ Total\ Savings\ (EUR)$$

Since environmental improvement is often connected with high investments[652] but may repay only in the long-term, it may be of particular interest for the managers to see the total savings (8)[653] resulting from environmental innovations, investments and activities. This is directly related to the total costs for the different environmental issue areas. As the case of Interface, Inc. for example proved, as early as the beginning stages, high savings are possible and should be tracked.[654]

$$(9)\ Share\ of\ Trained\ Staff\ (\%) = \frac{Total\ Number\ of\ Trained\ Staff}{Total\ Number\ of\ Staff}\ x\ 100$$

Management should not only be concerned with costs but also with the corporate culture and its support by its employees. As the environmental strategy should not only be superficial but deeply ingrained and lived by managers as well as by employees, they first have to understand it. Therefore, it is important to track the training progress of staff in environmental issues (9).

$$(10)\ Share\ of\ Green\ Customers\ (\%) = \frac{Total\ of\ Green\ Customers}{Total\ Customers}\ x\ 100$$

Besides, management should also be informed about the current public attitude concerning green products and involvement in environmental issues. The share of green customers (10) shows their acceptance and commitment.[655]

$$(11)\ Supply\ Chain\ Traffic\ Volume\ (km\ or\ tkm)$$

This indicator (11) measures the total distance/traffic volume of goods (or services) which have been driven between suppliers and the buyer, and between the buyer and customers.[656] As global SCs are widespread today and the miles a good is travelling make up large distances[657], resulting in immense environmental impacts, this indicator

652 See Walley (1994), p. 46.

653 See Rao (2005), p. 909; Rao (2010), p. 11.

654 See Anderson (1998), pp. 11, 16.

655 This indicator can be measured in sales unit (EUR). See subsection 6.4.8.

656 See ICAEW (URL): 'Supply Chain Miles'.

657 See Maria (2010), p. 1013.

should also be provided to the high-level managers, as some strategic changes concerning locations of sites, outsourcing and the thoughts about decentralization and centralization may be necessary to improve this indicator.

(12) *Supplier Environmental Sustainability Index*

As suppliers play a significant role in the environmental sustainability of the buyer company and as they have been tracked with the help of indicators as well, this information should also be provided to the managers. However, the manager should only see aggregated information and not each single indicator for each supplier. Therefore, the 'supplier environmental sustainability' index will be proposed (12).[658] It represents a "multi-item measure" and the total environmental performance of all suppliers. The indicator includes all suppliers' indicators and transforms them into one indicator with the help of adequate values and weight coefficients.[659] In order to receive this indicator the following two steps shall be followed, as is also presented with examples in figure 2 and figure 3.

Figure 2: Calculating the Supplier Environmental Sustainability Index (Step 1)

Step 1: Assigning a value for each environmental indicator for each supplier and calculating the mean value for each supplier.

Table 1:
Ex.: Supplier X

Environmental Criterion / Indicator	Value (V)
A	3
B	2
C	5
D	2
E	3
…	…
Sum *(S=A+B+C+D+E)*	15
Mean *(M=S/Number of Criteria)*	**3**

For example: A=Energy; B= Water; C=Material; D=Waste; E=Emissions

Source: Own representation and calculation.

The first step is to decide for each criterion/indicator and for each supplier, how he has performed. Each supplier will receive a value (V) ranging from 0 (very poor environmental performance) to 5 (very good environmental performance) for each

[658] See ICAEW (URL), incl. following sentence and quotation.

[659] See Tyteca (1996), p. 283; the following thoughts and calculation are i.a derived from Moneim (2013); the following will only be an exemplary presentation of how this indicator can be derived. However, it may be reasonable to change it to the own situation.

criterion.[660] The sum of the values for each criterion will be divided by the number of criteria. The resulting mean value for each supplier will be needed for the second step. Table 1 presents an example, where the mean value represents 3, showing a moderate performance of this supplier.

Figure 3: Calculating the Supplier Environmental Sustainability Index (Step 2)

Step 2: Weighting each supplier and calculating the final value for the indicator.

Table 2:
Ex.: All suppliers – Most important supplier with the worst environmental performance.

Suppliers	Value (V)	Weight (*w*)	V**w*
X GmbH	1	0,65	0,65
XY AG	5	0,1	0,5
Y KG	5	0,1	0,5
YZ GmbH	5	0,05	0,25
Z	5	0,1	0,5
...			
Sum		1	**2,4**

Final Value: Min:0, Max:5

Table 3:
Ex.: All suppliers – Most important supplier with the best environmental performance.

Suppliers	Value (V)	Weight (*w*)	V**w*
X GmbH	5	0,65	3,25
XY AG	2	0,1	0,2
Y KG	2	0,1	0,2
YZ GmbH	1	0,05	0,05
Z	2	0,1	0,2
...			
Sum		1	**3,9**

Final Value: Min:0, Max:5

Source: Own representation and calculation.

In step two each supplier will be listed with the respective mean value which will be weighted (*w*) according to its significance. This means that the value of each supplier will be multiplied by the weight decided for this particular supplier (V**w*). The weight reflects the importance of the supplier in terms of its mass fraction of material supplied to the buyer company. This weighting is extremely important for the calculation, as it

660 As the company may have a large number of suppliers, the evaluation of each supplier may exceed the benefits of this evaluation, and therefore it may be preferred to only evaluate the most significant suppliers in terms of environmental impact, meaning the suppliers with the highest mass fraction of input for the buyer company.

may be possible that the supplier which provides the firm with the most resources might be the one with the highest negative environmental impact and if all suppliers are accounted for equally, this will not show up in the final indicator. If it is addressed through weighting, a rather negative evaluation will result, although all others suppliers may be ranked best as shown in figure 3 table 2. On the other hand, if the most important supplier is very sophisticated in the area of environmental protection while the others are not, this will not show up either if no weights will be used. Yet, with the help of weight coefficients the significant supplier can outweigh the bad results of the other suppliers as one can see in figure 3 table 3. In the end the weighted values of all suppliers will be summed up. This value will range between 0 (very poor environmental performance) and 5 (very good environmental performance), representing the total environmental performance of all suppliers. If the value is unsatisfactory, one can dig deeper and look for the reason, i.e. which suppliers are responsible for the poor performance and in which environmental aspect. Certainly, this approach is a rather simple and arbitrary one, and very subjective concerning the values for each criterion but will still provide a way to present an indicator for the environmental performance of suppliers to executives.[661]

$$(13)\ \textit{Degree of Target Attainment}\ (\%) = \frac{\textit{Number of Achieved Environmental Targets}}{\textit{Number of Set Environmental Targets}}\ x\ 100$$

Finally, the management should always have an overview about the achievement of set targets. Especially, in case of environmental objectives and goals, the management should precisely track the attainment of those. However, in order to be able to track this indicator (13), targets have to be set and defined beforehand. A possible target might be the reduction of GHG emissions by 5% in the reference period. The indicators have, of course, to be adapted to the set targets, or if the company first starts to track general measures, such as the six basic indicators (1) – (6), it may decide afterwards about its concrete targets.

Since all these indicators would number more than twenty, including the intensity, and reduction indicators, as well as the other additionally proposed indicators, such as the shares, and therefore too much for showing top management, they have to be differentiated. The intensities, reduction and share indicators are only thought as supplement and should be applied if they are seen as very critical or essential for the firm. As it has already been mentioned, an entirely clear statement about which

[661] The company may also be able for future evaluations to decide for clear environmental objectives for the suppliers and may replace the value range from 0 until 5 by percentages according to the fulfilment of the objectives. For these percentages a value can then be assigned again (e.g. ranging from 0–5 for 0% –100%) and the average can be built. Concerning the relative weights it is also possible to make use of Analytical Hierarchy Process; see Moneim (2013).

indicators should be assessed for the corporate level cannot be made as it depends on the firm, sector, industry etc. Therefore, some additional indicators and advices are given in order to support the decision-making. However, the final decision should be made according to the firms' environmental strategy, which could on the other hand also be partly derived from the main indicators' information as the management can see the main problems of the firm, also called 'hot spots'[662]. Nevertheless, it is important that each corporate environmental indicator can be further explained by low-level indicators. This is necessary if one key indicator is critical and has to be pursued more precisely, which should be possible with the KEIs stated above in connection with the divisional indicators presented in section 6.4.

6.6 Overall Environmental Indicator

After defining a set of KEIs the question may arise whether it is possible to show the firm's total environmental performance with one single environmental indicator, in the same way that financial accounting has only one indicator that is often assumed as the most useful measure – ROI (Return on Investment).[663] However, the use of ROI as a measure is also often criticized, sometimes even seen as totally flawed.[664]

The questions which should be posed are: is an overall environmental indicator possible, and if yes, is it reasonable and useful? This section will not provide a concrete overall environmental indicator, as its added value is assumed to be still very questionable. The problems but also the opportunities, as well as a selection of proposed indicators or rather approaches will be presented.

Only limited research has been found for this topic, which may indicate that the development of an overall environmental indicator is not sufficiently valuable, reasonable or still too difficult and complex.

In order to be beneficial, the overall indicator should present the overall environmental performance of a company and facilitate the comparison between companies or industries.[665] Moreover, if a firm is only looking at a large number of indicators instead of a single indicator, it may disregard the whole and not gain an understanding of the wider context.[666] On the other hand, an overall environmental indicator provides additional challenges compared to an economic one, such as ROI, as it comprises many different aspects, with various metrics, and is not unified in monetary terms,

---

662 See Schröder (2014), p. 260f; chapter 7.

663 See Jacobson (1987), p. 470; Reece (1978), p. 28f.

664 See Jacobson (1987), p. 470.

665 See Jennen (1976), p. 152: Refers to the macro-level; Tyteca (1996), p. 282.

666 See Wiedmann (2010), p. 1651.

which is the case in the economic context.[667] Besides, it may indicate the wrong direction due to oversimplification.[668]

Nevertheless, there are some attempts to develop such an overall indicator, in particular in terms of a sustainability indicator, comprising all three dimensions. Gladen offers one example with the 'Sustainable Value Added' concerning the total resource efficiency.[669] Moneim et al. developed an overall sustainability index for measuring the sustainability of a manufacturing facility by making use of analytical hierarchy process (AHP).[670] For the aggregation of the different sustainability indicators into one, the indicators have to be normalized and are dimensionless.[671] The calculation is realized with weights received from AHP, where the results are three indicators for each dimension, thus also showing how to create one single environmental indicator.[672] Another approach is the arrangement of different indicators in a matrix where each dimension comprises different areas, such as horizontal indicators resembling the respective pollutant in diverse environmental media, and the vertical indicators referring to all pollutants in one media.[673] Another approach is the application of data envelopment analysis (DEA), by making use of quantities between 0 and 1, referring to the level of environmental performance.[674] Although this approach may be similar to the weighting approaches, it also makes use of "the ideas of the productive efficiency theory."[675] This approach also sees aggregate indicators resulting in measures without units or dimensions.[676] Xerox is a practical example of a firm which has created the Xerox Environmental Performance Metric that "provides a single indicator of environmental performance, thus uniting all functional elements within the supply chain."[677]

Another field, which may be worthwhile to explore in this context, is the family of footprints. These include i.a. the Ecological Footprint (EF), the Carbon Footprint (CF), the Water Footprint (WF) and the Material Footprint (MF).[678] The concept of the EF has been widely discussed and a clear opinion has not resulted yet, especially at the

667 See Jennen (1976), p. 151.
668 See Weidema (2008), p. 3f.
669 See Gladen (2014), p. 173f.
670 See Moneim (2013); for another index approach see Färe (2004).
671 See Moneim (2013), p. 3.
672 See Moneim (2013), pp. 4–12.
673 See Jennen (1976), p. 157ff.
674 See Tyteca (1996), pp. 281, 284, 289-296.
675 Tyteca (1996), p. 283.
676 See Tyteca (1996), p. 284.
677 McIntyre (1998), p. 150, incl. whole sentence. Although it is stated that there is one indicator, the article comprises several indicators and does not directly refer to a single one.
678 See Herva (2011); Gaussin (2013); Weidema (2008); Wiedmann (2013); Wiedmann (2010); Hoekstra (2006).

corporate level, as these indicators are often applied on a macro-level.[679] It may be only useful when applied together with other indicators and not as a single indicator as it may not include all environmental issues.[680] The EF concept has become obsolete in comparison to the CF.[681] Although the CF includes only one sphere of environmental issues, meaning the emissions resulting in global warming, it became very popular. It may be an initial step towards an overall environmental indicator but it is still not sufficient.[682] However, it is widely used, also for eco-labelling showing the whole carbon footprint for each product.[683] Practical examples include PepsiCo's brand Tropicana, the French Casino group, or Walkers Crisps. Despite its wide acceptance which may be due to its ease of calculation and understanding, it focuses only on global warming and emissions, which are certainly some of the most critical issues, but at the expense of other important issues such as waste water or chemicals.[684] This is the same with the WF and MF that comprise only specific aspects, and are therefore not sufficient for an overall environmental indicator.[685]

It can be concluded that an overall environmental indicator is still in its infancy and practitioners have to think about whether the effort of developing such an indicator is beneficial, as this highly aggregated indicator will lose information, may be too much in order to be useful and to show valuable information about the environmental situation of the firm. Thus, for internal purposes this indicator may not be as interesting as for public comparisons between companies, but this field will need further research in order to become really valuable and widespread.

679 See Herva (2011), p. 1688; Wiedmann (2010), p. 1646.

680 See Wiedmann (2010), pp. 1645, 1647.

681 See Wiedmann (2010), p. 1650, incl. following sentence.

682 See Gaussin (2013), p. 515.

683 See Gaussin (2013), p. 516; The Economist (URL); Martin (URL); Groupe Casino (URL), incl. following sentence.

684 See Weidema (2008), p. 3f; Gaussin (2013), p. 517; Wiedmann (2010), p. 1650.

685 See Herva (2011), pp. 1688, 1690; Wiedmann (2013), p. 6271; Wiedmann (2010), p. 1650; Hoekstra (2006).

7 Integration of Indicators into an Exemplary Framework for Implementation and Application: The 'Environmental Tree-Model'

This chapter will provide a proposal for detecting the most important environmental aspects, and offer a selection of possibilities that exist in implementing the selected indicators into the company, and how a possible model may look ('Environmental Tree-Model'), serving as a summary of all stated indicators at the same time. It will not give a detailed explanation of a model or system for the integration and application of indicators, as this would exceed the scope of this paper.

With the help of a LCA the firm is able to execute a "hot spot analysis"[686]. It helps to transfer the environmental strategy into practice by analysing the critical areas of activity that have the greatest impact on the firms' environmental sustainability and should be focused on with regards to indicators and initiatives. Nevertheless, it is also important to take action in other areas as long as they promise results in a fast and easy manner. The following matrix in figure 4 is an example of what a hot spot analysis may look like. All different value chain units will be regarded according to all important environmental media. If information has been collected in the different matrix fields, and some have given the alarm, the respective fields will be declared as hot spots, and the management will know immediately where the critical areas lie to this point of time.

Figure 4: An Example of a Hot Spot Analysis Matrix

	Suppliers	Procurement & Inbound Logistics	Technology Development	Firm Infrastructure	HRM	Operations	Outbound Logistics	Marketing, Sales & Service
Energy						Hot Spot		
Water								
Material								
Waste								
Emissions							Hot Spot	
Env. Costs			Hot Spot					
Fines								
Env. Awareness	Hot Spot							
Target Attainment								

Source: Own representation based on Schröder (2014), p. 261.

This might just be seen as an additional support tool for a special indicator model, as this presents only a very rough overview, or it may also be used as an indicator model by putting the respective indicators in the left column. It may serve before the final selection of indicators, to detect which environmental areas are of important interest after the LCA has been realized. Thus, it may be reasonable to start with the most

686 Schröder (2014), p. 260f, incl. whole sentence and following sentences.

important aspects in the firm before the firm is overstrained with the mass of improvement possibilities and does not start at all – 'better small than never'.

A concrete strategic system for the implementation and application of indicators in a wider context would be the well-known balanced scorecard (BSC)[687]. In the context of environmental indicators, the BSC might be of high usefulness, especially if the firm has already established one and is familiar with its usage.[688] As for some senior executives environmental sustainability is still not relevant and as they want to see the benefits of everything, the BSC may serve as such a tool as it would link the different perspectives, especially the financial, with the environmental aspects, and the managers may well align their business to an environmental perspective and goals.[689] This might be of specific importance because many firms do not use any strategic tools to link their environmental strategies to their corporate strategies, and therefore it often remains the concern of the single environmental departments but will not be linked to the financial objectives of the firm.[690] To integrate the environment strategically into the existent BSC, there are different approaches. It is possible to integrate key environmental indicators in each, some or one (mostly internal business process) of the four existing perspectives of the BSC, or to generate a new fifth perspective.[691] Another possibility is to create an extra sustainability balanced scorecard which may also be generated by each organizational level and division, becoming more aggregated in higher levels.[692] However, as some companies may still not have a BSC or want to track the environmental sustainability separately to conventional indicators or the economic situation, there are alternatives to the BSC.[693] Another model is the French Tableau de Bord which presents only the corporate key indicators (comparable to a dashboard of a car and also often presented like this).[694] Another visualization form presents the traffic light model (Red-Amber-Green), which is also often used in the context of the BSC.[695] It shows the target attainment level of each indicator.[696]

Therefore, the following so-called 'Environmental Tree – Model' has been developed, which can be regarded in figure 5.

---

[687] The balanced scorecard will not be explained in detail as this would again exceed the scope of the paper. For further explanations see i.a. Kaplan (1996).

[688] See Czymmek (2001), p. 23.

[689] See Epstein (2001), p. 3f; Hockerts (2001), p. 15f.

[690] See Hockerts (2001), p. 15.

[691] See Epstein (2001), p. 4; Czymmek (2001), p. 23; Hockerts (2001), p. 16.

[692] See Hockerts (2001), p. 17; Gladen (2014), p. 419.

[693] See Hockerts (2001), p. 16.

[694] See Franceschini (2007), p. 127f.

[695] See Kühnapfel (2014), p. 19; Colsman (2013), p. 64; Neely (2006), p. 11.

[696] See Neely (2006), p. 11.

Figure 5: The 'Environmental Tree-Model'

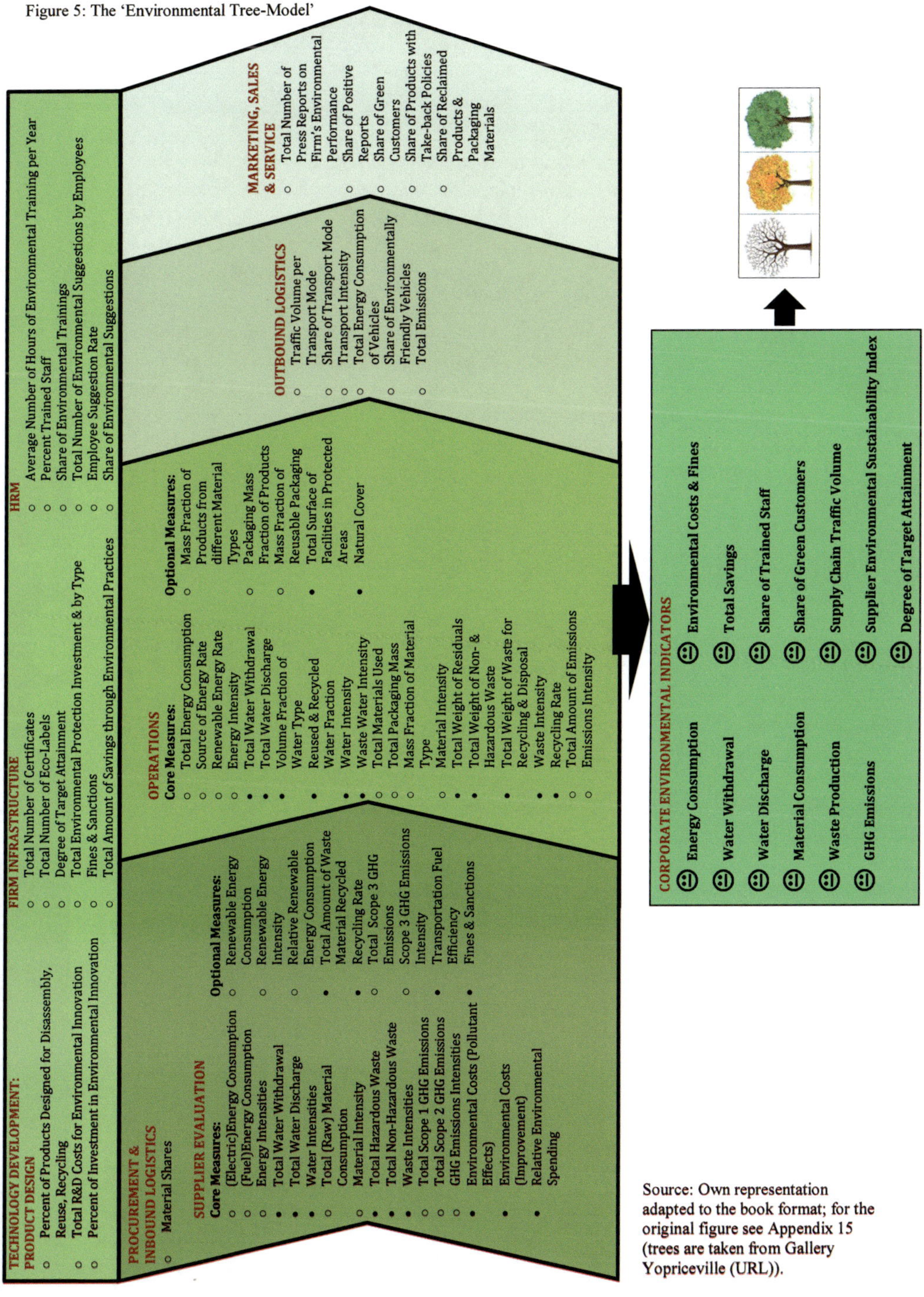

Source: Own representation adapted to the book format; for the original figure see Appendix 15 (trees are taken from Gallery Yopriceville (URL)).

It presents all stated value chain indicators in the first level, aggregated into the corporate key indicators, and resulting in the display of the environmental tree. In figure 5 all three levels are shown in order to get a general overview of all environmental indicators, but in practice it may be reasonable to present only the corporate level indicators with the 'environmental tree' to the top management. All these indicators will be assigned to a specific sign, being it a traffic light or an emoticon, like in figure 5. This sign will show the current achievement of the set target for this indicator.[697] If it is far from the target and no longer reachable the emoticon will be sad or the traffic light might be red, if the target achievement is in danger but can still be achieved, the emoticon may look like it is in figure 5. If the target may likely be reached and the firm is on a good track the emoticon may smile or the traffic light will be green. These results will be summed up and may be shown as a tree. According to its health (without leaves = very unhealthy; yellow leaves = not fully healthy but just sufficient; green leaves = perfectly healthy), one can see how the firm is doing respectively. The level of health of the environmental tree may be determined with the help of a predefined weighting system, depending on the number of emoticons that are sad, ok or happy or the traffic light being green, yellow or red. With the help of such an indicator model the management is able to see with one view the state of the firm in terms of its environmental performance and initiatives, and can dig deeper if necessary, in case that it is not doing as well as expected. Yet, this model is a distinct consideration of environmental indicators from conventional indicators. At the best, all indicators should be regarded as a whole and should be displayed in relation. Therefore, the BSC might be a good tool in order to implement and integrate the environmental indicators in the pre-existing indicator system.

This chapter is intended to give an impetus of how an indicator system may look and what the firm can do with the developed set of indicators. However, as always, each firm should implement the indicators adapted to its own situation, means and strategy in order to be successful and to create an environmentally sustainable organization. To achieve this, the environmental accounting has to work closely with all other departments to be able to create a valuable tool.[698]

---

[697] One corporate indicator represents the 'degree of target attainment'. This indicator might be unnecessary in the case of using such a model that shows for each indicator its target achievement with the help of a sign, such as the traffic light or emoticon and the overall target attainment is shown by the respective tree. However, the indicator was included, as this figure is also used as a summary in this case.

[698] See IFAC (2005), p. 30.

8 Conclusion

8.1 Critical Evaluation and Limitations

It became apparent that sustainability can no longer be a buzzword but has to be taken seriously in all aspects of life, especially in the context of business. In order to operationalize and integrate the concept of environmental sustainability in an organization, accounting can and should make its contribution. Possible effective operational tools which can support this effort and development are environmental indicators.[699] They should be integrated in a systematic way, as it was the case in this paper by following the structure of Porter's value chain, being applied to all internal value chain units but also going beyond by comprising suppliers, in particular from developing countries. Although there are already vast numbers of environmental indicators, such as those from the GRI guidelines or the ISO 14031 standard, and numerous studies, most of them have not been deemed as sufficiently precise and detailed concerning the indicators. They often do not offer the indicator formula, even though such visualization may help to understand the indicators better and faster, and they often lack clear definitions, as well as measurement units. Moreover, these guidelines present a huge number of indicators but do not provide any support for selection despite the fact that the amount of indicators would be too high to integrate them all. This paper tried to overcome these issues by offering a pre-defined and limited selection of generic environmental indicators, which aimed to be sufficiently explained in a comprehensible way by offering concrete formulas. Besides, the environmental indicators which are presented in literature are often only categorized according to their respective environmental media. This paper went one step further by assigning the different indicators to the respective value chain units and aggregating them to corporate environmental indicators as well.

However, in spite of the fact that environmental indicators are a useful tool, they present many obstacles and challenges which have to be faced and attempted to overcome.[700] The usefulness of the indicators vastly depends on their selection and the availability of crucial data. Thereby, not only the amount of indicators is decisive but especially whether they lead to the desired objective. Concerning the data, it is not only advantageous if the required data is already accessible but their accuracy is particularly significant. In case of estimations, the indicators and their interpretation cannot be exact. In addition, environmental indicators are generally backward-looking[701]. Although the overall goal might be to compare the created indicators with other companies, this may present a greater obstacle, as competitors may hide the

---

[699] See Zirkler (2011), p. 2.

[700] See Sommer (2010), p. 361f, incl. following sentences.

[701] Lagging indicators; see Hockerts (2001), p. 15.

respective data, or use different methodologies and possess different products and production processes, which make a valuable comparison rather unlikely. The mere addition of indicators in order to cover more environmental issues is not always correct as the added costs may exceed the added value, therefore their value should be always evaluated.[702] Lastly, as indicators (systems) present an aggregation of information this will automatically lead to a loss of information, which might be considerable.[703]

Since the environmental indicators have also been regarded in terms of suppliers from DCs, this presents additional challenges to the buyer company. Not only the different context, culture, understanding and language of DC suppliers may be problematic, but also the need to build trust and a profound relationship.[704] Measuring and monitoring suppliers is certainly important, but the distance may be an obstacle and beyond, the mere control of suppliers will not be sufficient in order to create an effective relationship.[705]

Apart from these rather general issues concerning environmental indicators, there are some concrete limitations for this bachelor dissertation which should be considered when making use of the presented environmental indicators. First of all, one has always to think of the consequences of one's actions. This means that even intendedly environmentally friendly actions could be thrown into reverse since those actions may have environmental impacts as well.[706] One example is renewable energy sources which also have negative effects on the environment. Thus, it would be wrong to increase the total energy consumption as long as it consists of renewable energy. The overall reduction of energy consumption should still be the main goal. However, despite the impacts of renewable energy on the environment there are those which assume that it is possible to turn them into advantages and even help the environment.[707] This aspect is mentioned in order to remind firms to consider all the consequences indicators may cause, and not only consider the positive effects but also the possibly negative side effects. The second limitation is the sole focus on the environmental dimension of the TBL. In order to create a truly and fully sustainable organization, not only the environmental sustainability but also the economic and social sustainability have to be included equally.[708] In this paper environmental indicators have been considered, excluding social and for the most part economic indicators, or indicators comprising two or all three dimensions, such as eco-efficiency

---

702 See Czymmek (2001), p. 14.
703 See Czymmek (2001), p. 14; Sommer (2010), p. 361.
704 See Rao (2002), pp. 634, 651; Hervani (2005), pp. 330, 346.
705 See Rao (2008), p. 24; Khan (2014), p. 1220; Robins (1997); Hervani (2005), p. 330.
706 See Ananthaswamy (2012), pp. 35, 38; UCS (URL), incl. following sentence.
707 See Ananthaswamy (2012), pp. 35, 38.
708 See Schröder (2014), p. 258.

indicators linking the economic and environmental dimension[709]. Especially social issues are often neglected in current indicator frameworks and need more attention in future.[710] However, with social indicators the problem of measurability may be even greater than with environmental indicators, although some do exist.[711] Moreover, ECIs have been excluded, as they were seen as rather difficult to implement,[712] however, they may also be important to measure the firm's impact on the local environment.[713] The third problematic aspect is the effort to create generic environmental indicators, trying to comprise all kinds and sizes of industries and companies. Thereby, the service industry may be a bit neglected[714] and the indicators may be better applied to manufacturing firms. Yet, an attempt has been made to offer some additional explanations for service firms as well. Nevertheless, the call for industry specific indicators exists and is certainly justified.[715] For instance, emissions indicator may be more important to the metal industry than to the service industry, and yet a service industry also produces emissions through business travels or through the usage of their offices. Therefore, the proposed indicators have their justification for all, especially with regards to comparative purposes and for a convenient access to the topic of greening the whole company. Concerning the different sizes of companies, SMEs have not been considered separately from large companies, although they predominate in Germany.[716] Their conditions, means and priorities may be different to those of large companies. However, this should not excuse their responsibility for the environment, even though they may have less impact. Despite their possibly limited budget for environmental actions they should be still able to build environmental indicators or a system, since some data may already be available and just has to be used. Nevertheless, they should evaluate the possible benefits of additional indicators precisely and may start on a smaller basis than large companies, but start nonetheless.[717] Due to the required application to all industries and companies, it has not been possible to define the indicators in perfect detail or to point out all possible issues or measures, as they have to be adapted accordingly to the firm's specific situation. One should not forget that some terms will mean something different to different firms and the nominators' and denominators' metrics have to be adapted as well. For instance, for the service industry the production output may be service hours

709 See subsection 5.4.1.

710 See Krajnc (2003), p. 281; Elkington (1999), p. 70.

711 See Schwarz (2002), p. 59; for social indicators i.a. Moneim (2013), pp. 3, 5, 7f; Hutchins (2008).

712 Only a small number of companies used ECIs: Bennett (1999).

713 See Hoffmann (2011), p. 164; DIN (2013), pp. 65–70: Also presenting examples of ECIs.

714 However, one must also consider that service firms usually have less impact on the environment than manufacturing firms, and therefore a slight prioritization might be justified.

715 See Zirkler (2011), p. 3, incl. following sentence.

716 See DESTATIS (URL).

717 See Schaltegger (2000a), p. 303.

while a manufacturing company will use the mass of products, thus the indicators and their respective measures have to be defined very precisely that everybody within the company will at least understand it equally. A more detailed description of indicators and all possible issues would also have exceeded the scope because a certain distinction between the specific industries would have been necessary. This dissertation aimed to offer a tool in order to create a completely environmentally sustainable company, therefore including all stages of the value chain, also encompassing the suppliers and the reclaim of products. Yet, it neglected the use stage of the products, which is certainly of high importance concerning the environmental impact.[718] If one is thinking for example of a washing machine which is making use of high amounts of water and energy during its life, it has significant effects on the environment. However, in this case it may be disputable whether the environmental impacts from the working life of products should be attributed to the companies or whether customers should be blamed directly.[719] As this is questionable and since it would have exceeded the scope of this work as well, the customer stage has not been included. The next critical point has been the selection of the indicators. The choice of indicators was subjective, as it was made according to the author's own thoughts of their usefulness and applicability, as no practical samples, no concrete research models or analyses have been used. Yet, a vast literature research has been conducted that included all aspects, and such a selection process will always be imbued of some degree of subjectivity.[720]

With reference to the stated limitations of this research, it is proposed that further research should especially be executed in the field of social indicators as well as sustainable indicators concerning the customer stage. The next step concerning the work of this paper would be to apply the selected indicators in practice to prove their applicability and usefulness.

8.2 Opportunities and Outlook

These limitations and issues demonstrate the challenges concerning this topic and may also present some reasons for its weak diffusion in accounting so far.[721] Nevertheless, the environmental issues are abundant and their devastating existence can no longer be denied or avoided; environmental accounting is unavoidable if firms want to stay competitive and survive in future[722]. Although it might not be easy to integrate

718 See McIntyre (1998), p. 153f; Schröder (2014), p. 258, incl. following sentence.

719 However, firms could be made responsible for designing the products in a way that they need fewer resources, such as less water for a washing machine; see Hoffmann (2011), p. 139f.

720 See Veleva (2001), p. 532.

721 See i.a. Burritt (2014), p. 330.

722 See Stubenrauch (2014), p. 12.

environmental issues in accounting, it offers benefits, as ignorance may lead to a competitive disadvantage[723] in the long-term, with others taking the lead. The engagement in this highly urgent subject is not only profitable for the company but will protect the environment and therefore secure the life on earth. Yet, the industry itself can certainly not do this alone. There is need for support of customers, buying environmentally friendly products and reconsidering their own (consumption) habits and waste production; investors, attaching importance on environmental and social aspects; and especially of the governments, supporting this development with the help of adapted eco-taxes, emissions certificates, laws and regulations. This will further put pressure on the companies and motivate them to act in a sustainable manner, and therefore enhance the general environmental situation. With the help of stricter regulations and laws, as well as better standards in terms of indicators, their implementation becomes easier. The indicators' importance is not negligible, which is confirmed by different quotations, as already Galileo Galilei stated that one should "Measure what is measurable, and make measurable what is not so"[724] because "If you can't measure it, you can't manage it"[725]. This provides a further case for the implementation of environmental indicators in accounting. Yet, the question which may arise is whether the accounting departments do have the corresponding capacities and means in order to develop such environmental indicators or systems.[726] It is well known that these units are often very busy and only have limited resources. Therefore, the demand for standardized instruments is high in order to implement such tools in a fast and easy manner. In general, it can be said that with its already high presence and increasing development of this topic in scientific work, efforts will be made in order to enhance standardization.

All this may sound as if environmental indicators and systems have not been applied so far, but this is not the case either. Numerous SMEs and large companies, especially active in environmental protection, already make use of these tools in order to evaluate, control and communicate their environmental performance and activities.[727] The only problem is that the controllers are often excluded from this effort of which the separate environmental departments are responsible.[728] As it has already been stated the interest in sustainability levelled off somehow concerning the controllers.[729] Nevertheless, the topic is still highly relevant and also the controllers will not be able to avoid tackling it. Therefore, it is assumed that the proper implementation of

723 Ibid.

724 I.a. Hallward (2008), p. 339; Schwarz (2007), p. 164.

725 I.a. Peccoud (2014); Faber (2014); Lisoski (2003).

726 See Zirkler (2011), p. 3, incl. following sentences.

727 See Sommer (2010), p. 362.

728 See Hockerts (2001), p. 15; see section 5.2.

729 See Weber (2015), p. 17, incl. following sentence.

environmental accounting is only a question of time, but as soon as it appears in a wider scope, it may be of interest whether environmental accounting will develop as a separate department, a subdivision of the conventional accounting or as a subordinate of the environmental department. Interface, Inc. for example, tracks its metrics at all BUs or manufacturing sites by financial, purchasing and sustainability departments, being also reported to the Corporate Sustainability department.[730] Hansgrohe SE sustainability accounting receives its environmental information company-wide from the EHS (Environment, Health, Safety) department.[731] The environmental accounting is subordinated to the corporate accounting but works closely together with the EHS.

Although this topic has been discussed over a seemingly long period, one should not forget that such fundamental corporate changes will need time, like the net present value which needed more than 30 years until it has become established in the US.[732] In order to integrate a completely new field into a firm and such a complex one[733], even more time may be required. Yet, accounting should not stop working on its establishment although it may still have to learn a lot, material flow cost accounting is also considered as a promising future tool in this field[734].

Firms are going already in the right direction when creating environmental reports for external purposes but the internal reporting has to follow now. Controllers as business partners of management have a responsibility to actively shape the development of corporate sustainable strategies and to integrate environmental aspects in their processes, instruments and structures, in order to successfully control the firm's environmental performance.[735] Moreover, not only the economic internal pressure for environmental engagement increases but especially the outside pressure coming from society and governments.[736] This is confirmed by future studies stating that the relevance of sustainable management will further increase. Especially, policy will try to further internalize external costs.

Besides all these stated reasons to implement sustainability in accounting, it should also be tackled in order to prove U Thant wrong, showing him that humans have finally understood and accepted their responsibility, stopped talking and started to use their genius and skills to protect the Earth, and consequently their future generations.

[730] See Hay (E-Mail).
[731] See Hansgrohe SE (E-Mail), incl. following sentence.
[732] See Weber (2015), p. 17.
[733] See Burritt (2014), pp. 327–329; Colsman (2013), p. 45.
[734] See Christ (2014), p. 1f; Burritt (2014), p. 335.
[735] See Berlin (2014), p. 47.
[736] See Wiesehahn (2015), p. 29, incl. following sentences.

Appendices

Appendix 1: Geographic Distribution of Sustainability Science Publications

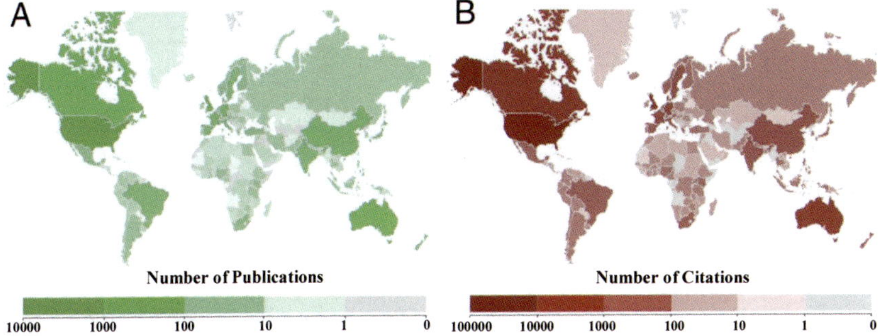

Source: Bettencourt (URL).

Appendix 2: Overall Increase in Publications on Corporate Sustainability

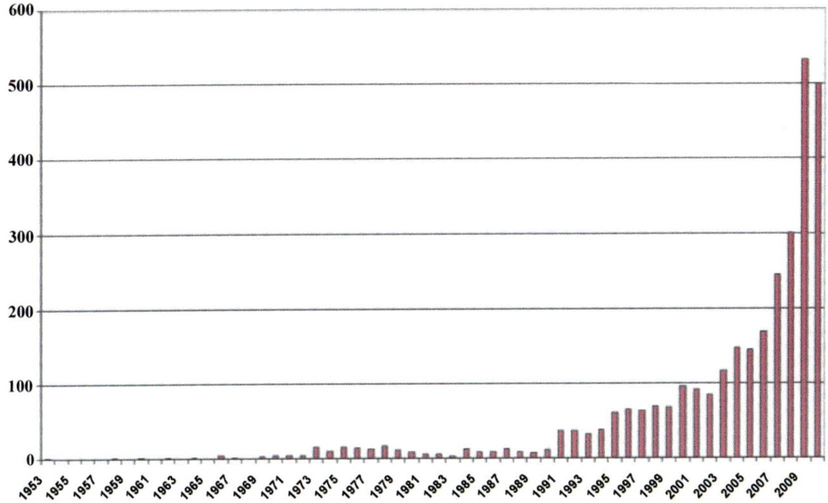

Fig. 1. Number of publications on corporate sustainability per year 1953-2010.

Source: Linnenluecke (2012), p. 385.

Appendix 3: Value Creation with the Help of Sustainability

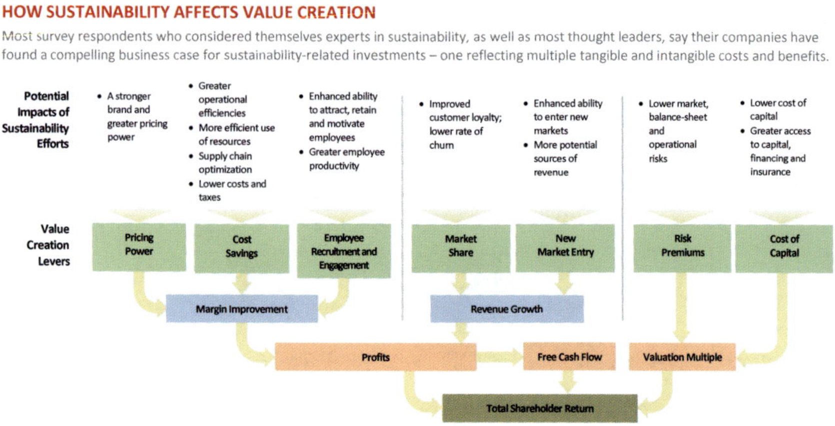

Source: Berns (2009), p. 6.

Appendix 4: Four Approaches to Environmental Accounting

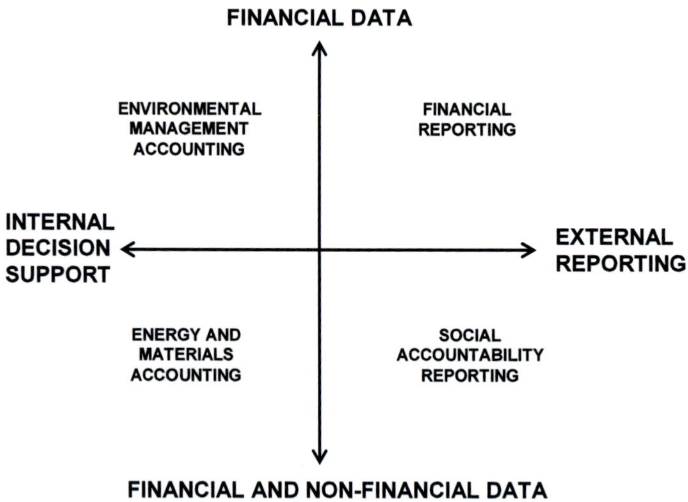

Source: Bartolomeo (2000), p. 33.

Appendix 5: Different Categories of Environmental Accounting

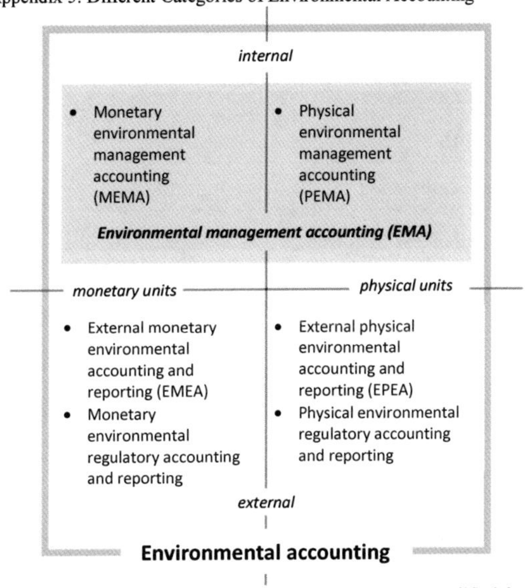

Source: Schaltegger (2000b), p. 16; Burritt (2002a), p. 27 modified from Bartolomeo (2000), p. 33.

Appendix 6: The Value System

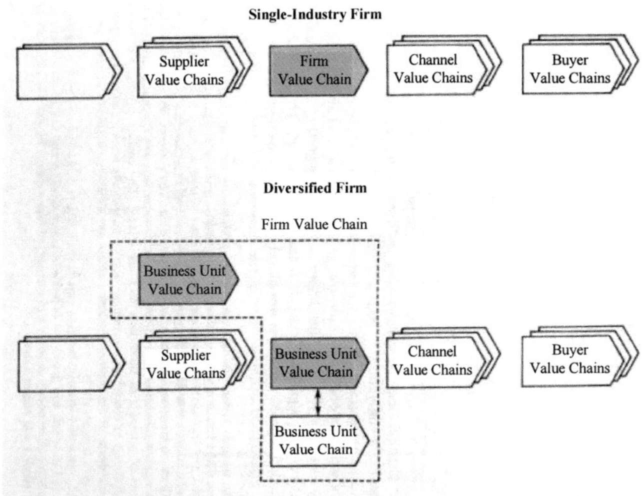

Figure 2-1. The Value System

Source: Porter (1985), p. 35.

Appendix 7: Porter's Generic Value Chain

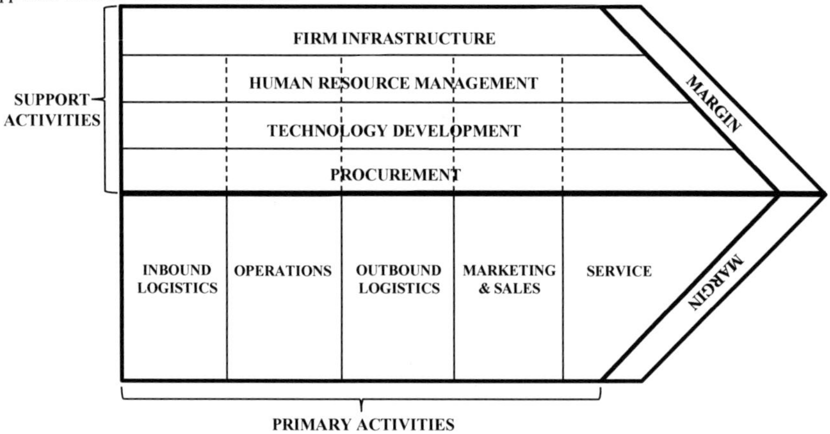

Source: Self-created drawing based on Porter (1985), p.37.

Appendix 8: Overview of a Company's Processes and Operations

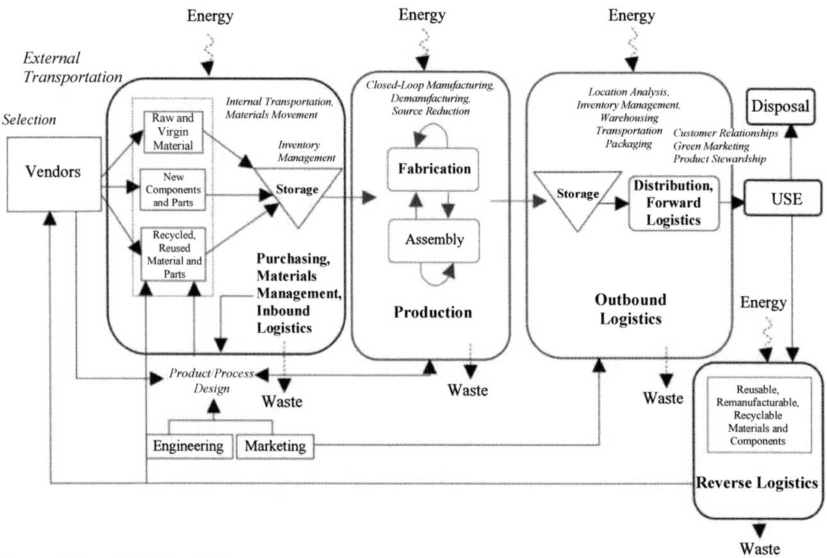

Source: Hervani (2005), p. 335.

Appendix 9: Operations of an Organization

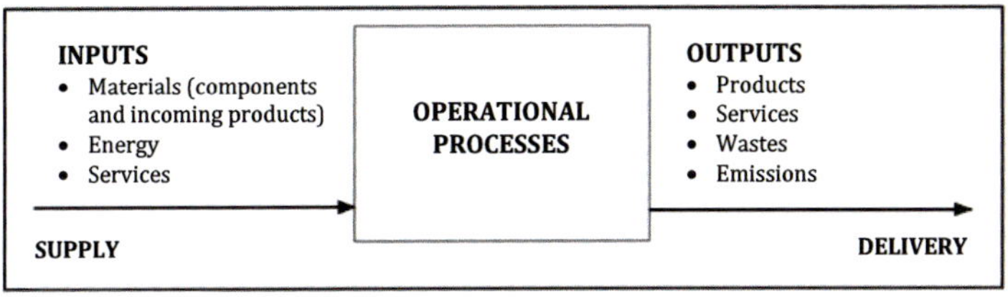

Figure 3 - Operations of the organization (general overview)

Source: DIN (2013), p. 24.

Appendix 10: Inputs and Outputs of Manufacturing

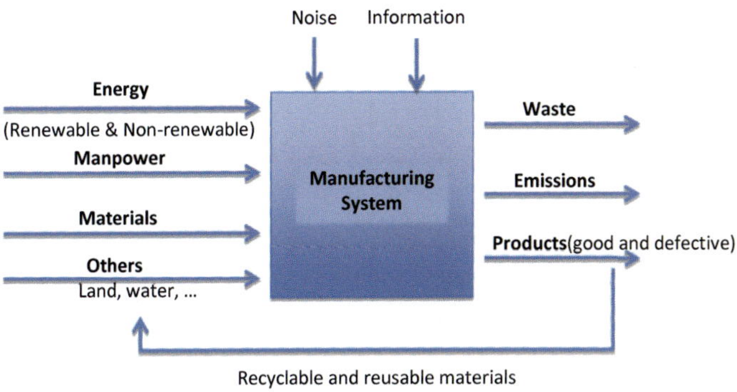

Figure 1 – Manufacturing System inputs and outputs

Source: Moneim (2013), p. 3.

Appendix 11: Conversion Table

Anmerkung: x(y) bedeutet x·10^y

Anmerkung: Für alle Stoffe wurden die [unteren] Heizwerte verwendet, nicht die [oberen Heizwerte] Brennwerte

						Wasserstoff			Diesel		Stein-kohle	Methan (Erdgas)		Rohöl	
		J	kWh	cal	BTU	kg	Nm^3	l LH_2	kg	l	kg	kg	Nm^3	kg	Barrel
	J	1	2,778(-7)	0,2388	9,478(-4)	8,335(-9)	9,274(-8)	1,177(-7)	2,328(-8)	2,799(-8)	3,412(-8)	1,802(-8)	2,511(-8)	2,388(-8)	1,751(-10)
	kWh	3,6(6)	1	8,598(5)	3412	3,001(-2)	0,3339	0,4239	8,380(-2)	0,1008	0,1228	6,487(-2)	9,041(-2)	8,598(-2)	6,304(-4)
	cal	4,187	1,163(-6)	1	3,968(-3)	3,490(-8)	3,883(-7)	4,930(-7)	9,746(-8)	1,172(-7)	1,429(-7)	7,544(-8)	1,051(-7)	1(-7)	7,331(-10)
	BTU	1055	2,931(-4)	252,0	1	8,795(-6)	9,785(-5)	1,242(-4)	2,456(-5)	2,953(-5)	3,600(-5)	1,901(-5)	2,650(-5)	2,520(-5)	1,848(-7)
Wasser-stoff	kg	1,200(8)	33,33	2,865(7)	1,137(5)	1	11,13	14,13	2,793	3,358	4,094	2,162	3,013	2,865	2,101(-2)
	Nm^3	1,078(7)	2,995	2,575(6)	1,022(4)	8,988(-2)	1	1,270	0,2510	0,3018	0,3679	0,1943	0,2708	0,2575	1,888(-3)
	l LH_2	8,493(6)	2,359	2,028(6)	8049	7,079(-2)	0,7876	1	0,1977	0,2377	0,2898	0,1530	0,2133	0,2028	1,487(-3)
Diesel	kg	4,296(7)	11,93	1,026(7)	4,072(4)	0,3581	3,984	5,058	1	1,202	1,466	0,7741	1,079	1,026	7,523(-3)
	l	3,573(7)	9,925	8,534(6)	3,386(4)	0,2978	3,314	4,207	0,8317	1	1,219	0,6438	0,8973	0,8534	6,257(-3)
Stein-kohle	kg	2,931(7)	8,141	7(6)	2,778(4)	0,2443	2,718	3,451	0,6822	0,8203	1	0,5281	0,7360	0,7	5,132(-3)
Methan (Erdgas)	kg	5,550(7)	15,42	1,326(7)	5,260(4)	0,4626	5,147	6,535	1,292	1,553	1,894	1	1,394	1,326	9,718(-3)
	Nm^3	3,982(7)	11,06	9,511(6)	3,774(4)	0,3319	3,693	4,689	0,9269	1,114	1,359	0,7175	1	0,9511	6,973(-3)
Rohöl	kg	4,187(7)	11,63	1(7)	3,968(4)	0,3490	3,883	4,930	0,9746	1,172	1,429	0,7544	1,051	1	7,331(-3)
	Barrel	5,711(9)	1,586(3)	1,364(9)	5,413(6)	47,60	529,6	672,4	132,9	159,8	194,9	102,9	143,4	136,4	1

Source: DWV (URL).

Appendix 12: Conversion into Gigajoules

Coal	**GJ**	**Crude Oil**	**GJ**	**Gasoline**	**GJ**	**Natural Gas**	**GJ**	**Electricity**	**GJ**
tonne (metric)	26,00	barrel	6,22	US gallon	0,125	therm	0,1055	kilowatt-hour	0,0036
ton (short)	23,59	tonne (metric)	44,80	tonne (metric)	44,80	1000 cubic feet	1,1046	megawatt-hour	3,6000
ton (long)	26,42	ton (short)	40,64	**Diesel**		1000 cubic meters	39,01	gigawatt-hour	3600,0
		ton (long)	45,52	US gallon	0,138	MMBtu	1,055		
				tonne (metric)	43,33				
				Fuel Oil					
				US gallon	0,144				
				tonne (metric)	40,19				

Source: GRI (2000-2011), p. 8.

Appendix 13: The 3 Emission Scopes

FIGURE 3. **Overview of scopes and emissions across a value chain**

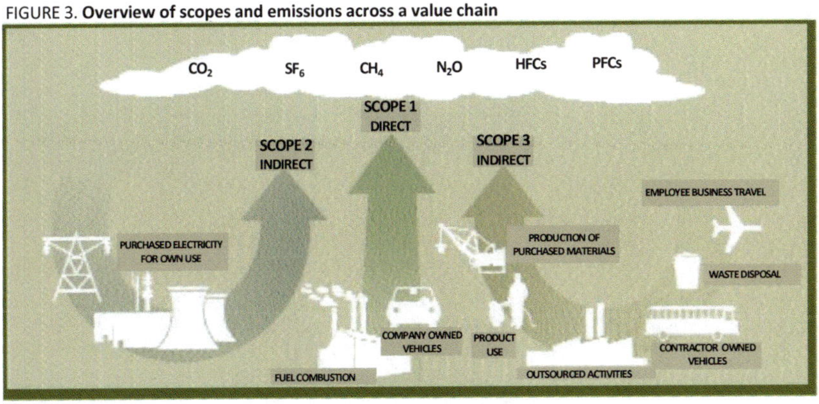

Source: WBCSD (2004), p. 26 adopted from NZBCSD (2002), p. 10.

Appendix 14: EcoMetrics of Interface

PDF 1:

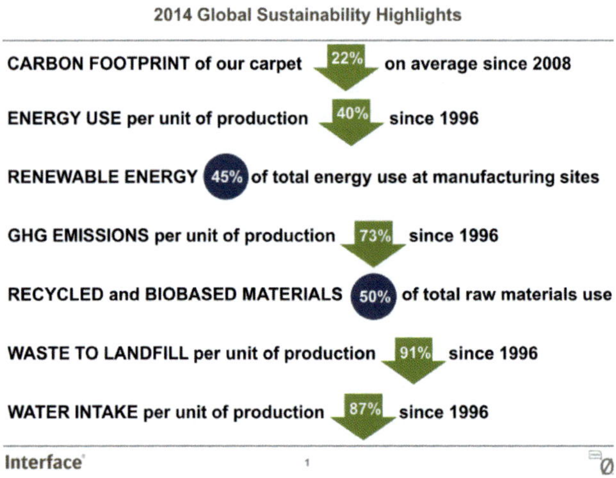

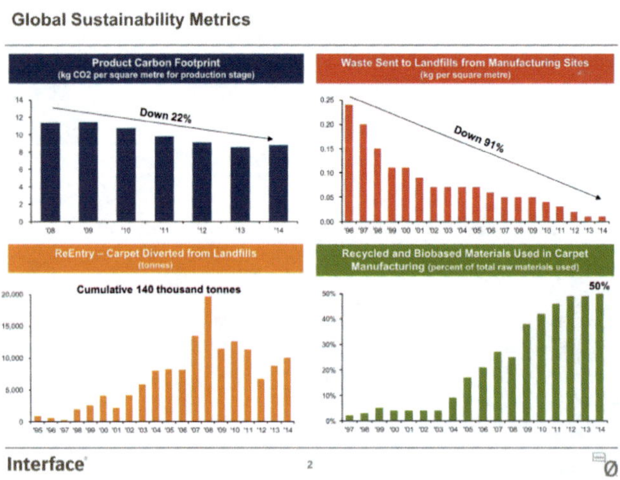

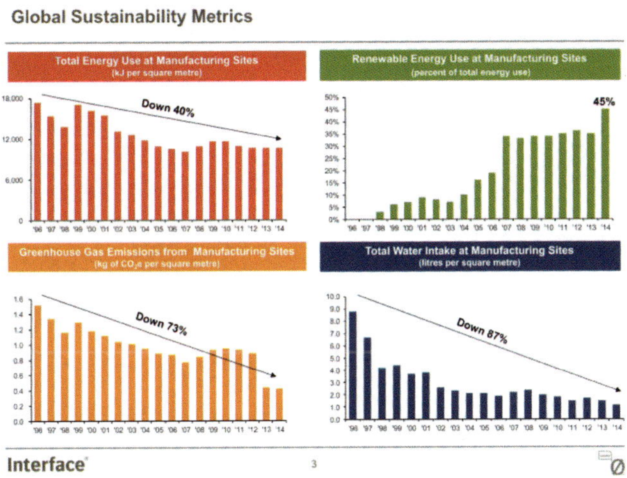
Global Sustainability Metrics
Total Energy Use at Manufacturing Sites
(kJ per square metre)
Down 40%
18,000
12,000
6,000
0
'96 '97 '98 '99 '00 '01 '02 '03 '04 '05 '06 '07 '08 '09 '10 '11 '12 '13 '14
Renewable Energy Use at Manufacturing Sites
(percent of total energy use)
45%
50%
45%
40%
35%
30%
25%
20%
15%
10%
5%
0%
'96 '97 '98 '99 '00 '01 '02 '03 '04 '05 '06 '07 '08 '09 '10 '11 '12 '13 '14
Greenhouse Gas Emissions from Manufacturing Sites
(kg of CO2e per square metre)
Down 73%
1.6
1.4
1.2
1.0
0.8
0.6
0.4
0.2
0.0
'96 '97 '98 '99 '00 '01 '02 '03 '04 '05 '06 '07 '08 '09 '10 '11 '12 '13 '14
Total Water Intake at Manufacturing Sites
(litres per square metre)
Down 87%
10.0
9.0
8.0
7.0
6.0
5.0
4.0
3.0
2.0
1.0
0.0
'96 '97 '98 '99 '00 '01 '02 '03 '04 '05 '06 '07 '08 '09 '10 '11 '12 '13 '14
Interface
3

Two relevant pages of PDF 2:

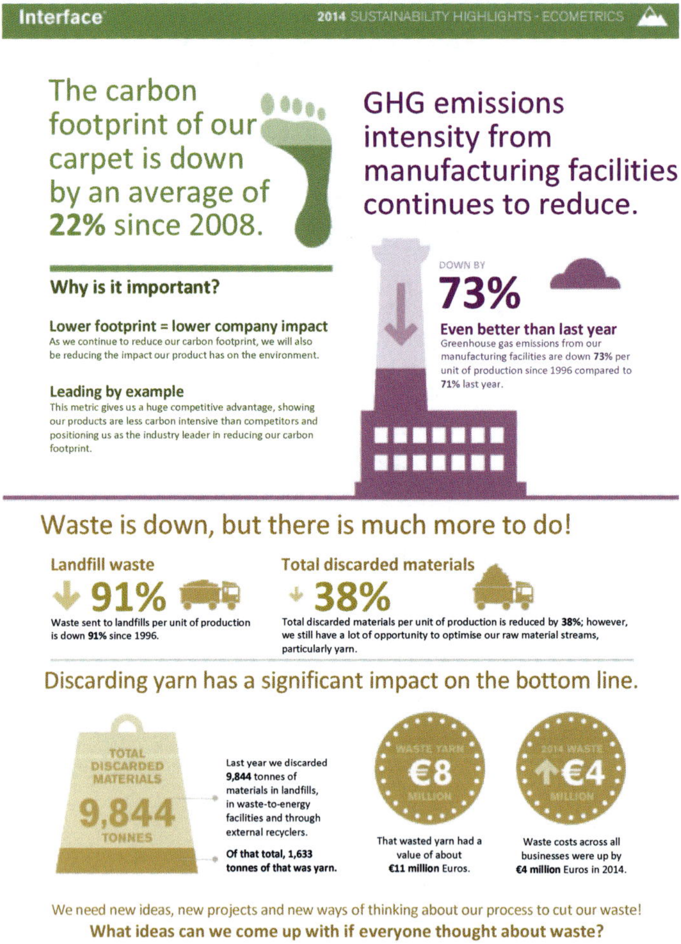

Interface® 2014 SUSTAINABILITY HIGHLIGHTS - ECOMETRICS

The carbon footprint of our carpet is down by an average of **22%** since 2008.

Why is it important?

Lower footprint = lower company impact
As we continue to reduce our carbon footprint, we will also be reducing the impact our product has on the environment.

Leading by example
This metric gives us a huge competitive advantage, showing our products are less carbon intensive than competitors and positioning us as the industry leader in reducing our carbon footprint.

GHG emissions intensity from manufacturing facilities continues to reduce.

DOWN BY
73%

Even better than last year
Greenhouse gas emissions from our manufacturing facilities are down **73%** per unit of production since 1996 compared to **71%** last year.

Waste is down, but there is much more to do!

Landfill waste
91%
Waste sent to landfills per unit of production is down **91%** since 1996.

Total discarded materials
38%
Total discarded materials per unit of production is reduced by **38%**; however, we still have a lot of opportunity to optimise our raw material streams, particularly yarn.

Discarding yarn has a significant impact on the bottom line.

TOTAL DISCARDED MATERIALS 9,844 TONNES

Last year we discarded **9,844** tonnes of materials in landfills, in waste-to-energy facilities and through external recyclers.

Of that total, 1,633 tonnes of that was yarn.

WASTE YARN €8 MILLION
That wasted yarn had a value of about **€11 million** Euros.

2014 WASTE €4 MILLION
Waste costs across all businesses were up by **€4 million** Euros in 2014.

We need new ideas, new projects and new ways of thinking about our process to cut our waste!
What ideas can we come up with if everyone thought about waste?

Interface **2014** SUSTAINABILITY HIGHLIGHTS - ECOMETRICS

Highlights

The following metrics are highlights of our recent sustainability progress that will help us better share our Interface story. To obtain the full metrics, visit the sustainability sections of Loop and www.interfaceglobal.com.

Use of renewable energy increases from **35** to **45%** at manufacturing facilities.

Global

Today, 4 of our 7 manufacturing plants use 100% renewable electricity – Scherpenzeel, Craigavon, LaGrange and West Point.

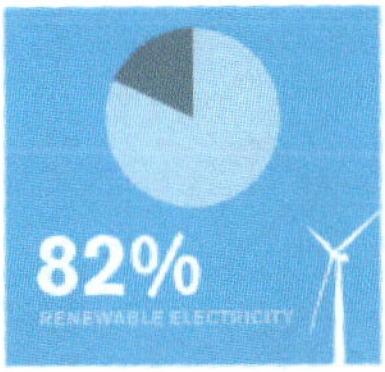

Global

82% of the electricity used at our manufacturing facilities is from renewable sources.

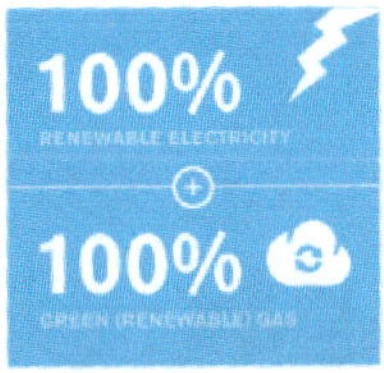

Scherpenzeel

By securing green gas for our Scherpenzeel facility, our European team enabled us to achieve both 100% renewable electricity and 100% renewable gas, making it the first Interface facility to have this distinction. Furthermore, this resulted in a 98% reduction in GHG emissions for our European manufacturing facilities since 1996.

Recycled and biobased materials now make up **50%** of total raw materials used.

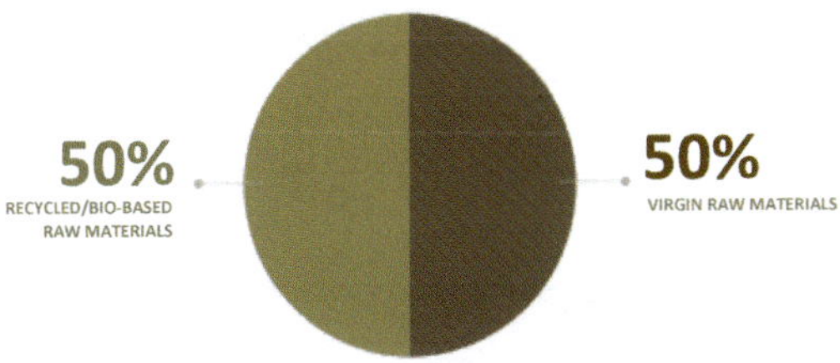

Source: Hay (E-Mail).

Appendix 15: The Original 'Environmental Tree-Model'

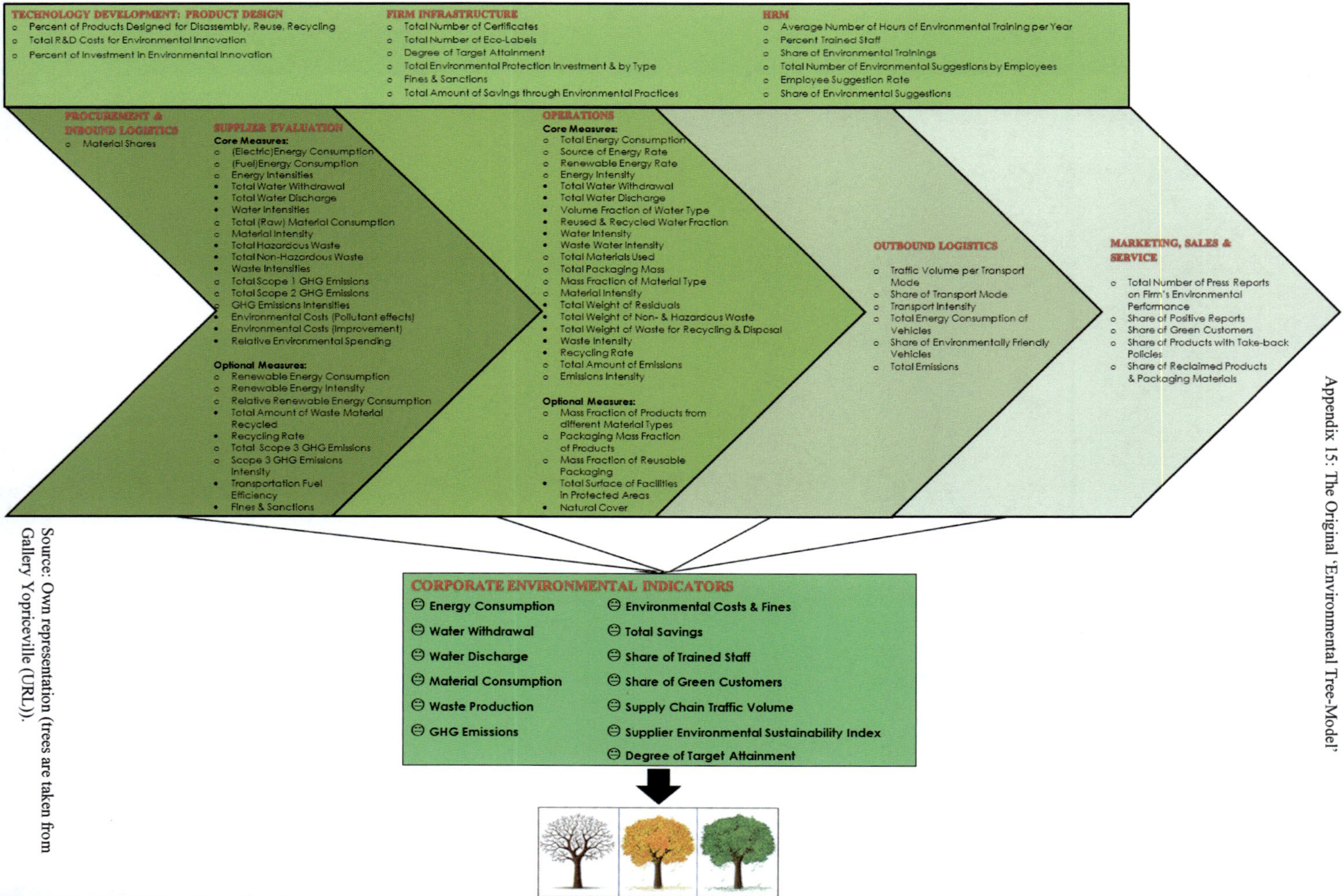

Source: Own representation (trees are taken from Gallery Yopriceville (URL)).

References

accenture strategy and First Carbon Solutions (2015): Supply Chain Sustainability Revealed: A Country Comparison. Supply Chain Report 2014-15. CDP (Ed.). URL: https://www.cdp.net/CDPResults/CDP-Supply-Chain-Report-2015.pdf.

Ahi, Payman and Searcy, Cory (2015a): An analysis of metrics used to measure performance in green and sustainable supply chains. In: Journal of Cleaner Production, Vol. 86, pp. 360–377. DOI: 10.1016/j.jclepro.2014.08.005. URL: http://www.sciencedirect.com/science/article/pii/S0959652614008270/pdfft?md5=26a7465ad7cbfa94204c3835e801f940&pid=1-s2.0-S0959652614008270-main.pdf.

Ahi, Payman and Searcy, Cory (2015b): Assessing sustainability in the supply chain: A triple bottom line approach. In: Applied Mathematical Modelling, Vol. 39 Iss. 10-11, pp. 2882–2896. DOI: 10.1016/j.apm.2014.10.055. URL: http://www.sciencedirect.com/science/article/pii/S0307904X14005575/pdfft?md5=a51c7025292c043b452320156830732d&pid=1-s2.0-S0307904X14005575-main.pdf.

Akamp, Marion and Müller, Martin (2013): Supplier management in developing countries. In: Journal of Cleaner Production, Vol. 56, pp. 54–62. DOI: 10.1016/j.jclepro.2011.11.069. URL: http://www.sciencedirect.com/science/article/pii/S0959652611005087/pdfft?md5=a825a6864cebfef2eb4260665549e6f7&pid=1-s2.0-S0959652611005087-main.pdf.

Alfthan, Björn and Hislop, Lawrence (2015): Rapid change in the Arctic. In: UNEP (Ed.): UNEP Year Book 2014. Emerging Issues in Our Global Environment: United Nations Pubns, pp. 60–65. URL: http://www.unep.org/yearbook/2014/PDF/UNEP_YearBook_2014.pdf.

Amini, Mehdi and Bienstock, Carol C. (2014): Corporate sustainability: an integrative definition and framework to evaluate corporate practice and guide academic research. In: Journal of Cleaner Production, Vol. 76, pp. 12–19. DOI: 10.1016/j.jclepro.2014.02.016. URL: http://www.sciencedirect.com/science/article/pii/S0959652614001607/pdfft?md5=cde921fade643868bbf35b3f780ccd5c&pid=1-s2.0-S0959652614001607-main.pdf.

Ananthaswamy, Anil and Le Page, Michael (2012): How clean is green? In: New Scientist, Vol. 213 Iss. 2849, pp. 34–38. URL: http://www.sciencedirect.com/science/article/pii/S0262407912602474/pdfft?md5=bfea9b19b4fe03e202b574b502917804&pid=1-s2.0-S0262407912602474-main.pdf.

Anderson, Ray C. (1998): Mid-Course Correction. Toward A Sustainable Enterprise : The Interface Model. Atlanta, Ga: Peregrinzilla Press.

Aras, Güler and Crowther, David (2009): Making sustainable development sustainable. In: Management Decision, Vol. 47 Iss. 6, pp. 975–988. DOI: 10.1108/00251740910966686. URL: http://www.emeraldinsight.com/doi/pdfplus/10.1108/00251740910966686.

Armenteras, Dolors and Finlayson, C. Max (2012): Biodiversity. Chapter 5. In: UNEP (Ed.): Global Environment Outlook GEO 5. Environment for the future we want. Nairobi, Kenya: United Nations Environment Programme, pp. 133–166. URL: http://www.unep.org/geo/pdfs/geo5/GEO5_report_full_en.pdf.

Arundel, Anthony; Kemp, René and Parto, Saeed (2006): Indicators for environmental innovation: what and how to measure. Chapter 21. In: Marinova, Dora; Annandale, David; Phillimore, John (Eds.): The international handbook on environmental technology management. Cheltenham, UK, Northampton, MA: Edward Elgar (Elgar original reference), pp. 324–339. URL: http://arno.unimaas.nl/show.cgi?fid=15710.

Arvidsson, Rickard et al. (2012): Energy use indicators in energy and life cycle assessments of biofuels: review and recommendations. In: Journal of Cleaner Production, Vol. 31, pp. 54–61. DOI: 10.1016/j.jclepro.2012.03.001. URL: http://www.sciencedirect.com/science/article/pii/S0959652612001345/pdfft?md5=80708d51efdfd725d9d13d8111210481&pid=1-s2.0-S0959652612001345-main.pdf.

Askham, Cecilia; Gade, Anne Lill and Hanssen, Ole Jørgen (2012): Combining REACH, environmental and economic performance indicators for strategic sustainable product development. In: Journal of Cleaner Production, Vol. 35, pp. 71–78. DOI: 10.1016/j.jclepro.2012.05.015. URL: http://www.sciencedirect.com/science/article/pii/S0959652612002387/pdfft?md5=5b58b96eb5df175acf077c3d033b4714&pid=1-s2.0-S0959652612002387-main.pdf.

Atkinson, Giles (2000): Measuring Corporate Sustainability. In: Journal of Environmental Planning and Management, Vol. 43 Iss. 2, pp. 235–252. DOI: 10.1080/09640560010694. URL: http://www.tandfonline.com/doi/pdf/10.1080/09640560010694.

Awasthi, Anjali; Chauhan, Satyaveer S. and Goyal, S.K (2010): A fuzzy multicriteria approach for evaluating environmental performance of suppliers. In: International Journal of Production Economics, Vol. 126 Iss. 2, pp. 370–378. DOI: 10.1016/j.ijpe.2010.04.029. URL: http://www.sciencedirect.com/science/article/pii/S0925527310001507/pdfft?md5=6e1b0bae9d6fbcdb18fb8ec0c72dafc7&pid=1-s2.0-S0925527310001507-main.pdf.

Azadi, Majid et al. (2015): A new fuzzy DEA model for evaluation of efficiency and effectiveness of suppliers in sustainable supply chain management context. In: Computers & Operations Research, Vol. 54, pp. 274–285. DOI: 10.1016/j.cor.2014.03.002. URL: http://www.sciencedirect.com/science/article/pii/S0305054814000513/pdfft?md5=d06d5d93c87417e8f51e59ef8945356c&pid=1-s2.0-S0305054814000513-main.pdf.

Azapagic, A. and Perdan, S. (2000): Indicators of Sustainable Development for Industry: A General Framework. In: Trans IChemE, Vol. 78 Part B, pp. 243–261. DOI: 10.1205/095758200530763. URL: https://www.escholar.manchester.ac.uk/api/datastream?publicationPid=uk-ac-man-scw:103501&datastreamId=POST-PEER-REVIEW-PUBLISHERS.PDF.

Bai, Chunguang and Sarkis, Joseph (2010): Integrating sustainability into supplier selection with grey system and rough set methodologies. In: International Journal of Production Economics, Vol. 124 Iss. 1, pp. 252–264. DOI: 10.1016/j.ijpe.2009.11.023. URL: http://www.sciencedirect.com/science/article/pii/S0925527309004265/pdfft?md5=7c055eb54642ce0ab3a20655d39bba1c&pid=1-s2.0-S0925527309004265-main.pdf.

Bai, Chunguang et al. (2012): Evaluating ecological sustainable performance measures for supply chain management. In: Supply Chain Management (Supply Chain Management: An International Journal), Vol. 17 Iss. 1, pp. 78–92. DOI: 10.1108/13598541211212221. URL: http://www.emeraldinsight.com/doi/pdfplus/10.1108/13598541211212221.

Barra, Ricardo; Portas, Pierre and Watkinson, Roy Victor (2012): Chemicals and Waste. Chapter 6. In: UNEP (Ed.): Global Environment Outlook GEO 5. Environment for the future we want. Nairobi, Kenya: United Nations Environment Programme, pp. 167–192. URL: http://www.unep.org/geo/pdfs/geo5/GEO5_report_full_en.pdf.

Bartolomeo, Matteo et al. (2000): Environmental management accounting in Europe: current practice and future potential. In: European Accounting Review, Vol. 9 Iss. 1, pp. 31–52. DOI: 10.1080/096381800407932. URL: http://web.b.ebscohost.com/ehost/viewarticle?data=dGJyMPPp44rp2%2fdV0%2bnjisfk5Ie46a9JrquzT7Ck63nn5Kx95uXxjL6trUqtqK5Js5a0Uq6uuEuylr9lpOrweezp33vy3%2b2G59q7Ra%2bst1CzqLNOrqmkhN%2fk5VXj5KR84LPgjOac8nnls79mpNfsVbKosEiwp7M%2b5OXwhd%2fqu4ji3MSN6uLSffbq&hid=118.

Bask, Anu and Kuula, Markku (2011): Measuring Supply Chain Level Environmental Sustainability - Case Nokia. In: International Journal of Business Insights & Transformation, Vol. 3 Iss. S3, pp. 16–24. URL: http://web.a.ebscohost.com/ehost/viewarticle?data=dGJyMPPp44rp2%2fdV0%2bnjisfk5Ie45PFIr6m2T7Kk63nn5Kx95uXxjO62tUevrK1KrqezOLSwski4qrU4v8OkjPDX7Ivf2fKB7eTnfLuntUiwrbdPsqe2PurX7H%2b72%2bw%2b4ti7evPepIzf3btZzJzfhrurtk%2b1p7VOtJzkh%2fDj34y75uJ%2bxOvqhNLb9owA&hid=4209.

Baskaran, Venkatesan; Nachiappan, Subramanian and Rahman, Shams (2012): Indian textile suppliers' sustainability evaluation using the grey approach. In: International Journal of Production Economics, Vol. 135 Iss. 2, pp. 647–658. DOI: 10.1016/j.ijpe.2011.06.012. URL: http://www.sciencedirect.com/science/article/pii/S0925527311002787/pdfft?md5=51c048287ff3e446d0737b0b3a432884&pid=1-s2.0-S0925527311002787-main.pdf.

Baumgartner, Rupert J. and Ebner, Daniela (2010): Corporate Sustainability Strategies: Sustainability Profiles and Maturity Levels. In: Sust. Dev. (Sustainable Development), Vol. 18 Iss. 2, pp. 76–89. DOI: 10.1002/sd.447. URL: http://sti.uem.mz/documentos/d_sustentavel/corporatesustainabilitystrategies.pdf.

Begon, Michael; Harper, John L. and Townsend, Colin R. (2009): Naturschutz. In: Townsend, Colin R.; Begon, Michael; Harper, John L. (Eds.): Ökologie. Berlin, Heidelberg: Springer Berlin Heidelberg, pp. 543–576.

Bell, Simon and Morse, Stephen (2008): Sustainability indicators. Measuring the immeasurable? 2nd ed. London, Sterling, VA: Earthscan.

Benedetto, Luca de and Klemeš, Jiří (2009): The Environmental Performance Strategy Map: an integrated LCA approach to support the strategic decision-making process. In: Journal of Cleaner Production, Vol. 17 Iss. 10, pp. 900–906. DOI: 10.1016/j.jclepro.2009.02.012. URL: http://www.sciencedirect.com/science/article/pii/S0959652609000316/pdfft?md5=ae26904c159df9bc53d6af21fd6200cb&pid=1-s2.0-S0959652609000316-main.pdf.

Bennett, Martin and James, Peter (1999): ISO 14031 and the Future of Environmental Performance Evaluation. In: Bennett, Martin; James, Peter; Klinkers, Leon (Eds.): Sustainable measures. Evaluation and reporting of environmental and social performance. Sheffield, UK: Greenleaf Pub., EBSCO HOST HTML Full Text, without page numbers. URL: http://web.a.ebscohost.com/ehost/viewarticle?data=dGJyMPPp44rp2%2fdV0%2bnjisfk5Ie45PFIr6m2T7Kk63nn5Kx95uXxjO62tUevra1KrqezOLCwr0u4q7A4v8OkjPDX7Ivf2fKB7eTnfLuptkuvqrVIsaqkhN%2fk5VXj5KR84LPgjeac8nnls79mpNfsVbesr021qaR%2b7ejrefKz5ozr4sSN6uLSffbq&hid=4209.

Bennett, Martin; Schaltegger, Stefan and Zvezdov, Dimitar (2013): Exploring Corporate Practices in Management Accounting for Sustainability. London: Institute for Chartered Accountants of England and Wales (ICAEW). URL: https://www.icaew.com/~/media/Files/Products/sustainability/tecpln11121a-exploring-corporate-practices-in-management-for-accounting.pdf.

Bergius, Susanne (URL): Im Kerngeschäft fehlt Nachhaltigkeit. Zwar verstärkt der deutsche Mittelstand sein soziales Engagement - doch ein verantwortliches Management des eigenen Betriebs bildet derzeit noch die Ausnahme. Studie. Handelsblatt. http://www.handelsblatt.com/unternehmen/mittelstand/studie-im-kerngeschaeft-fehlt-nachhaltigkeit/7334096.html (19.05.2015).

Berlin, Sebastian; Schulze, Mike and Stehle, Alexander (2014): Umsetzung eines Green Controllings: Integration als Erfolgsrezept. In: Controller Magazin, Iss. 6, pp. 47–49.

Berns, Maurice et al. (2009): The Business of Sustainability. What It Means to Managers Now. In: MIT Sloan Management Review, Vol. 51 Iss. 1, PDF pp. 2-8. URL: http://www.monitorpro.si/media/objave/dokumenti/2010/4/20/sas___the_business_of_sustainability.pdf.

Bettencourt, Luis M.A and Kaur, Jasleen (URL): Evolution and structure of sustainability science. PNAS. Proceeding of the National Academy of Sciences of the United States of America (Vol. 108 Iss. 49). http://www.pnas.org/content/108/49/19540/F2.expansion.html (05.04.2015).

BIPM (2006): The International Systems of Units (SI). Le système international d'unités. 8th ed. Sèvres: Organisation intergouvernementale de la Convention du Mètre. Bureau International des Poids et Mesures (BIPM). URL: http://www.bipm.org/utils/common/pdf/si_brochure_8_en.pdf.

Björklund, Maria; Martinsen, Uni and Abrahamsson, Mats (2012): Performance measurements in the greening of supply chains. In: Supply Chain Management: An International Journal, Vol. 17 Iss. 1, pp. 29–39. DOI: 10.1108/13598541211212186. URL: http://www.emeraldinsight.com/doi/pdfplus/10.1108/13598541211212186.

Boer, Yvo de and van Bergen, Barend (2012): Expect the Unexpected. Building business value in a changing world. Executive Summary. [S.l.]: KPMG International. URL: http://www.kpmg.com/Global/en/IssuesAndInsights/ArticlesPublications/Documents/building-business-value-exec-summary.pdf.

Bonini, Sheila and Swartz, Steven (2014): Profits with purpose: How organizing for sustainability can benefit the bottom line. McKinsey & Company (Ed.). URL: http://www.mckinsey.com/~/media/McKinsey/dotcom/client_service/ Sustainability/PDFs/McK%20on%20SRP%202014/ SRP_2014_Profits%20with%20Purpose.

Brown, Halina Szejnwald; Jong, Martin de and Lessidrenska, Teodorina (2009): The rise of the Global Reporting Initiative: a case of institutional entrepreneurship. In: Environmental Politics, Vol. 18 Iss. 2, PDF pp. 1-45. URL: http://www.ksg.harvard.edu/m-rcbg/CSRI/publications/ workingpaper_36_brown.pdf.

Brown, Mark Graham (1996): Keeping Score. Using the Right Metrics to Drive World-Class Performance. New York: Quality Resources. Distributed by AMACOM Books.

Burritt, Roger L.; Hahn, Tobias and Schaltegger, Stefan (2002a): An Integrative Framework of Environmental Management Accounting - Consolidating the Different Approaches of EMA into a Common Framework and Terminology. In: Bennett, Martin; Bouma, Jan Jaap; Wolters, Teun (Eds.): Environmental Management Accounting: Informational and Institutional Developments. Dordrecht [etc.]: Kluwer Academic Publishers (Eco-Efficiency in Industry and Science, 9), pp. 21–35.

Burritt, Roger L.; Hahn, Tobias and Schaltegger, Stefan (2002b): Towards a Comprehensive Framework for Environmental Management Accounting - Links Between Business Actors and EMA Tools. In: Australian Accounting Review, Vol. 12 Iss. 28, PDF pp. 1-24. DOI: 10.1111/ j.1835-2561.2002.tb00202.x. URL: http://www.apira2013.org/past/apira2001/ papers/Burritt110.pdf.

Burritt, Roger and Schaltegger, Stefan (2014): Accounting towards sustainability in production and supply chains. In: The British Accounting Review, Vol. 46 Iss. 4, pp. 327–343. DOI: 10.1016/j.bar.2014.10.001. URL: http://www.sciencedirect.com/science/article/pii/S0890838914000626/ pdfft?md5=131d6dfec20d36841438df7fa70f4987&pid=1-s2.0-S0890838914000626-main.pdf.

Burschel, Carlo (2001): Ökologisches Produktdesign. 4.2. In: Bundesumweltministerium, Umweltbundesamt (Eds.): Handbuch Umweltcontrolling. 2nd ed. München: Vahlen, pp. 269–280.

Büyüközkan, Gülçin and Çifçi, Gizem (2011): A novel fuzzy multi-criteria decision framework for sustainable supplier selection with incomplete information. In: Computers in Industry, Vol. 62 Iss. 2, pp. 164–174. DOI: 10.1016/j.compind.2010.10.009. URL: http://www.sciencedirect.com/science/article/pii/S0166361510001570/pdfft?md5=e23457cf3790e18ecafc7f6a71953b8c&pid=1-s2.0-S0166361510001570-main.pdf.

Campos, Lucila M.S. et al. (2015): Environmental performance indicators: a study on ISO 14001 certified companies. In: Journal of Cleaner Production, Vol. 99, pp. 286–296. DOI: 10.1016/j.jclepro.2015.03.019. URL: http://www.sciencedirect.com/science/article/pii/S095965261500236X/pdfft?md5=03babb90e14f58fcc5f129c7e1ed6973&pid=1-s2.0-S095965261500236X-main.pdf.

Caplice, Chris and Sheffi, Yossi (1994): A Review and Evaluation of Logistics Metrics. In: Int Jrnl Logistics Management (The International Journal of Logistics Management), Vol. 5 Iss. 2, pp. 11–28. DOI: 10.1108/09574099410805171. URL: http://web.mit.edu/sheffi/www/selectedMedia/rvw_eval_log_metrics_intl_jrnl_log_vol05_no02_pp11_28.pdf.

Carbon Decisions (2010): Carbon footprinting. An introduction for organisations. URL: http://www.carbondecisions.ie/resources/footprint_for_organisations.pdf.

Carter, Craig R.; Kale, Rahul and Grimm, Curtis M. (2000): Environmental purchasing and firm performance: an empirical investigation. In: Transportation Research Part E: Logistics and Transportation Review, Vol. 36 Iss. 3, pp. 219–228. DOI: 10.1016/S1366-5545(99)00034-4. URL: http://www.sciencedirect.com/science/article/pii/S1366554599000344/pdfft?md5=3765c8165d1ee6e1bb89b0da634be5dd&pid=1-s2.0-S1366554599000344-main.pdf.

Charter, M. et al. (2001): Supply Chain Strategy and Evaluation. The Center for Sustainable Design (The Sigma Project). URL: www.projectsigma.co.uk/RnDStreams/RD_supply_chain_strategy.pdf.

Chopra, Sunil and Meindl, Peter (2007): Supply Chain Management. Strategy, Planning, and Operation. 3rd ed. Upper Saddle River, N.J: Pearson Prentice Hall.

Christ, Katherine L. and Burritt, Roger L. (2014): Material flow cost accounting: a review and agenda for future research. In: Journal of Cleaner Production, PDF pp. 1-12. DOI: 10.1016/j.jclepro.2014.09.005. URL: http://www.sciencedirect.com/science/article/pii/S0959652614009354/pdfft?md5=c937010ac3a5cfabd732daaa854b5fa3&pid=1-s2.0-S0959652614009354-main.pdf.

Clift, Roland; Allenby, Brad and Ayres, Robert (2000): Forum on Sustainability. In: Clean Products and Processes, Vol. 2 Iss. 1. Springer-Verlag. pp. 67–70. DOI: 10.1007/s100980050052. URL: http://link.springer.com/content/pdf/10.1007%2Fs100980050052.pdf.

Colsman, Bernhard (2013): Nachhaltigkeitscontrolling. Strategien, Ziele, Umsetzung. Wiesbaden: Springer Fachmedien Wiesbaden; Imprint: Springer Gabler (SpringerLink : Bücher). URL: http://link.springer.com/content/pdf/10.1007%2F978-3-8349-3599-1.pdf.

Czymmek, Frank and Faßbender-Wynands, Ellen (2001): Die Bedeutung der Balanced Scorecard im Rahmen eines auf Kennzahlen basierenden Umwelt-Controlling. Die Balanced Scorecard im Umwelt-Controlling. Arbeitsbericht Nr.6. Univ.-Prof. Dr. Dr. h.c. Günter Beuermann (Ed.). Universität zu Köln. Köln. URL: http://www.econbiz.de/archiv/k/uk/soperationsr/bs_umwelt-controlling.pdf.

Dale, Virginia H. and Beyeler, Suzanne C. (2001): Challenges in the development and use of ecological indicators. In: Ecological Indicators, Vol. 1 Iss. 1, pp. 3–10. DOI: 10.1016/S1470-160X(01)00003-6. URL: http://www.sciencedirect.com/science/article/pii/S1470160X01000036/pdfft?md5=c8cd086336c157cd2ec71d1bb8c1cafb&pid=1-s2.0-S1470160X01000036-main.pdf.

Danciu, Victor (2013): The sustainable company. New challenges and strategies for more sustainability. In: Theoretical and applied economics : GAER review, Vol. XX Iss. 9, pp. 7–26. URL: http://store.ectap.ro/articole/898.pdf.

Danse, Myrtille and Wolters, Teun (2003): Sustainable Coffee in the Mainstream. The Case of the SUSCOF Consortium in Costa Rica. In: Greener Management International, Vol. 2003 Iss. 43, pp. 36–51. DOI: 10.9774/GLEAF.3062.2003.au.00006. URL: http://www.iscom.nl/publicaties/danseprf2.pdf.

Delhaes, D.; Kersting, S. and Ludwig, T. (2015): EU reformiert den Emissionshandel. Ab 2019 werden die Verschmutzungsrechte reduziert. Experten raten, den Verkehrssektor mit einzubeziehen. In: Handelsblatt, Iss. 87, p. 8f.

DESTATIS (URL): Mehr als 60% der tätigen Personen arbeiten in kleinen und mittleren Unternehmen. DESTATIS Statistisches Bundesamt. https://www.destatis.de/DE/ZahlenFakten/GesamtwirtschaftUmwelt/UnternehmenHandwerk/KleineMittlereUnternehmenMittelstand/Aktuell_.html (21.06.2015).

DIN (2013): Umweltmanagement - Umweltleitsungsbewertung - Leitlinien (ISO 14031:2013); Deutsche und Englische Fassung EN ISO 14031:2013. DIN EN ISO 14031. Berlin: Beuth Verlag.

Ditz, Daryl and Ranganathan, Janet (1997): Measuring Up. Toward a Common Framework for Tracking Corporate Environmental Performance. Washington, DC: World Resources Institute. URL: http://www.wri.org/sites/default/files/pdf/measuringup_bw.pdf.

Dobin, Diana (2009): Greenwashing Harms Entire Movement. In: Lodging Hospitality, Vol. 65 Iss. 14, p. 42. URL: http://web.a.ebscohost.com/ehost/viewarticle?data=dGJyMPPp44rp2%2fdV0%2bnjisfk5Ie45PFIr6m2T7Kk63nn5Kx95uXxjO62tEe2pbBIr6ueTbipt1Kzr55Zy5zyit%2fk8Xnh6ueH7N%2fiVauntEmwprVQtaqwPurX7H%2b72%2bw%2b4ti7evPepIzf3btZzJzfhruqs0iuqbBQsJzkh%2fDj34y75uJ%2bxOvqhNLb9owA&hid=4104.

Dobos, Imre and Vörösmarty, Gyöngyi (2014): Green supplier selection and evaluation using DEA-type composite indicators. In: International Journal of Production Economics, Vol. 157, pp. 273–278. DOI: 10.1016/j.ijpe.2014.09.026. URL: http://www.sciencedirect.com/science/article/pii/S0925527314003041/pdfft?md5=c1861f3c3c8173b4f712e48dfc80298e&pid=1-s2.0-S0925527314003041-main.pdf.

Doublet, Geneviève et al. (2014): Key Environmental Performance Indicators for a simplified LCA in food supply chains. In: Schenck, R.; Huizenga, D. (Eds.), Proceedings of the 9th International Conference on Life Cycle Assessment in the Agri-Food Sector (LCA Food 2014). San Francisco, USA, ACLCA, Vashon, WA, USA. PDF pp. 1-10. URL: http://www.senseproject.eu/images/publications/Oral246.pdf.

Durth, Rainer; Körner, Heiko and Michaelowa, Katharina (2002): Neue Entwicklungsökonomik. Stuttgart: Lucius & Lucius (UTB für Wissenschaft, 2306).

DWV (URL): Energietabelle für die Umrechnung verschiedener Energieeinheiten und -äquivalente. Deutscher Wasserstoff- und Brennstoffzellen-Verband e.V. Berlin. http://www.dwv-info.de/wissen/tabellen/wiss_enr.html (14.07.2015).

Dyllick, Thomas and Hockerts, Kai (2002): Beyond The Business Case For Corporate Sustainability. In: Bus. Strat. Env. (Business Strategy and the Environment), Vol. 11 Iss. 2, pp. 130–141. DOI: 10.1002/bse.323. URL: http://www.iwoe.unisg.ch/~/media/internet/content/dateien/instituteundcenters/iwoe/forschung%20nh/beyond%20the%20business%20case.pdf.

Eccles, Robert; Ioannou, Ioannis and Serafeim, George (URL): Is sustainability now the key to corporate success? The Guardian. http://www.theguardian.com/sustainable-business/sustainability-key-corporate-success (01.04.2015).

EIU (2010): Managing for sustainability. The Economist Intelligence Unit Limited (EIU). URL: http://graphics.eiu.com/upload/eb/Enel_Managing_for_sustainability_WEB.pdf.

Elkington, John (1994): Towards the sustainable corporation: win-win-win business strategies for sustainable development. In: California Maangement Review, Vol. 36 Iss. 2, pp. 90–100. URL: http://web.a.ebscohost.com/ehost/viewarticle?data=dGJyMPPp44rp2%2fdV0%2bnjisfk5Ie45PFIr6m2T7Kk63nn5Kx95uXxjO62tUewp61KrqezOLawrk%2b4p7A4v8OkjPDX7Ivf2fKB7eTnfLuqskmwrrZOta2khN%2fk5VXj5KR84LPgjeac8nnls79mpNfsVbeqr0iwp7FRsaikfu3o63nys%2b585LzzhOrK45Dy&hid=4212.

Elkington, John (1999): Cannibals With Forks. The Triple Bottom Line Of 21st Century Business. Oxford: Capstone.

Elkington, John (2004): Enter the Triple Bottom Line. Chapter 1. In: Henriques, Adrian; Richardson, Julie (Eds.): The Triple Bottom Line: Does It All Add Up? Assessing the sustainability of business and CSR. London, Sterling, VA: Earthscan, pp. 1–16. URL: http://kmhassociates.ca/resources/1/Triple%20Bottom%20Line%20a%20history%201961-2001.pdf.

Elliott, Jennifer A. (2006): An Introduction to Sustainable Development. 3rd ed. Abingdon, Oxon, New York: Routledge, Taylor & Francis Group (Routledge perspectives on development).

Ellram, Lisa M. (1990): The Supplier Selection Decision in Strategic Partnerships. In: Journal of Purchasing and Materials Management, Vol. 26 Iss. 3, pp. 8–14.

Enarsson, Leif (1998): Evaluation of suppliers: how to consider the environment. In: Int Jnl Phys Dist & Log Manage (International Journal of Physical Distribution & Logistics Management), Vol. 28 Iss. 1, pp. 5–17. URL: http://www.emeraldinsight.com/doi/pdfplus/10.1108/09600039810205935.

Engelman, Robert (2013): Beyond Sustainababble. Chapter 1. In: Assadourian, Erik; Prugh, Tom; Starke, Linda (Eds.): Is Sustainability Still Possible? State of the World 2013. Washington, DC [etc.]: Island Press (State of the world, 2013), pp. 3–16.

Epstein, Marc J. (2008a): Implementing Corporate Sustainability: Measuring and Managing Social and Environmental Impacts. In: Strategic Finance, pp. 26–31. URL: http://www.imanet.org/pdfs/public/sf/2008_01/01-08-epstein.pdf.

Epstein, Marc J. (2008b): Making Sustainability Work. Best Practices in Managing and Measuring Corporate Social, Environmental and Economic Impacts. 1st ed. Sheffield, UK, San Francisco: Greenleaf Pub.; Berrett-Koehler Publishers.

Epstein, Marc J. and Roy, Marie-Josée (2003): Improving Sustainability Performance: Specifying, Implementing and Measuring Key Principles. In: Journal of General Management, Vol. 29 Iss. 1, pp. 15–31. URL: http://web.b.ebscohost.com/ehost/viewarticle?data=dGJyMPPp44rp2%2fdV0%2bnjisfk5Ie45PFIr6m2T7Kk63nn5Kx95uXxjO62tUeypbBIr6ueULirs1Kwpp5Zy5zyit%2fk8Xnh6ueH7N%2fiVa%2bts0%2buqa5KsK%2bkhN%2fk5VXj5KR84LPgjeac8nnls79mpNfsVa%2bnsEywqrFMpNztiuvX8lXu2uRe8%2bLqbOPu8gAA&hid=118.

Epstein, Marc J. and Wisner, Priscilla S. (2001): Good Neighbors: Implementing Social and Environmental Strategies with the BSC. Balanced Scorecard Report. Insight, Experience & Ideas For Strategy-Focused Organizations. Harvard Business School Publishing (Ed.). URL: http://www.greenprof.org/wp-content/uploads/2010/02/Sustainability-BSC.pdf.

Erol, Ismail; Sencer, Safiye and Sari, Ramazan (2011): A new fuzzy multi-criteria framework for measuring sustainability performance of a supply chain. In: Ecological Economics, Vol. 70 Iss. 6, pp. 1088–1100. DOI: 10.1016/j.ecolecon.2011.01.001. URL: http://www.sciencedirect.com/science/article/pii/S0921800911000024/pdfft?md5=ec90b4c308da29ad5b45bb4091cb0998&pid=1-s2.0-S0921800911000024-main.pdf.

Faber, Véronique et al. (2014): 'If you can't measure it, you can't manage it': microinsurance by the numbers. In: Enterprise Development and Microfinance, Vol. 25 Iss. 1, pp. 41–56. DOI: 10.3362/1755-1986.2014.005. URL: http://web.a.ebscohost.com/ehost/viewarticle?data=dGJyMPPp44rp2%2fdV0%2bnjisfk5Ie45PFIr6m2T7Kk63nn5Kx95uXxjO62tUewpq1KrqezOLewsk64qq84v8OkjPDX7Ivf2fKB7eTnfLujr0%2b0r69Ot6uySKTi34bls%2bOGpNrgVeDr5j7y1%2bVVv8Skeeyzt02vprNOtKqkfu3o63nys%2b585LzzhOrK45Dy&hid=4204.

Färe, R.; Grosskopf, S. and Hernandez-Sancho, F. (2004): Environmental performance: an index number approach. In: Resource and Energy Economics, Vol. 26 Iss. 4, pp. 343–352. DOI: 10.1016/j.reseneeco.2003.10.003. URL: http://www.sciencedirect.com/science/article/pii/S0928765503000794/pdfft?md5=54b30ac8740fed26e1b97f1a143f3501&pid=1-s2.0-S0928765503000794-main.pdf.

Flämig, Heike (2001): Verkehr und Logistik. 5.2. In: Bundesumweltministerium, Umweltbundesamt (Eds.): Handbuch Umweltcontrolling. 2nd ed. München: Vahlen, pp. 357–374.

FOCUS online (URL): Tornado fegte mit bis zu 250 Stundenkilometern über Bayern hinweg. Wetter aktuell. FOCUS. http://www.focus.de/panorama/wetter-aktuell/wetter-aktuell-schwere-unwetter-zwei-schwerverletzte-in-freiburg_id_4682060.html (25.06.2015).

Formentini, Marco and Taticchi, Paolo (2014): Corporate sustainability approaches and governance mechanisms in sustainable supply chain management. In: Journal of Cleaner Production, PDF pp 1-14. DOI: 10.1016/j.jclepro.2014.12.072. URL: http://www.sciencedirect.com/science/article/pii/S0959652614013699/pdfft?md5=c8eea1bb21d0dd98b16a44260674a366&pid=1-s2.0-S0959652614013699-main.pdf.

Franceschini, Fiorenzo; Galetto, Maurizio and Maisano, Domenico (2007): Management by Measurement. Designing Key Indicators and Performance Measurement Systems. With 87 Figures and 62 Tables. Berlin [u.a.]: Springer. URL: http://link.springer.com/content/pdf/10.1007%2F978-3-540-73212-9.pdf.

Freeman, R. Edward (2010a): Strategic management. A stakeholder approach. Cambridge: Cambridge University Press.

Freeman, R. Edward et al. (2010b): Stakeholder Theory. The State Of The Art. Cambridge, New York: Cambridge University Press.

Freeman, R. Edward and McVea, John (2001): A Stakeholder Approach to Strategic Management. Darden Business School Working Paper No. 01-02. PDF pp. 1-32. URL: http://papers.ssrn.com/paper.taf?abstract_id=263511.

Friedman, Milton (2007): The Social Responsibility of Business is to Increase Its Profits. In: Zimmerli, Walther Ch.; Richter, Klaus; Holzinger; Markus (Eds.): Corporate ethics and corporate governance. Berlin, New York: Springer, pp. 173–178. URL: http://link.springer.com/content/pdf/10.1007%2F978-3-540-70818-6_14.pdf.

Gaddis, Erica Brown et al. (2012): Water. Chapter 4. In: UNEP (Ed.): Global Environment Outlook GEO 5. Environment for the future we want. Nairobi, Kenya: United Nations Environment Programme, pp. 97–132. URL: http://www.unep.org/geo/pdfs/geo5/GEO5_report_full_en.pdf.

Gallego-Álvarez, Isabel and Vicente-Villardón, José Luis (2012): Analysis of environmental indicators in international companies by applying the logistic biplot. In: Ecological Indicators, Vol. 23, pp. 250–261. DOI: 10.1016/j.ecolind.2012.03.024. URL: http://www.sciencedirect.com/science/article/pii/S1470160X1200132X/pdfft?md5=70e2fc8ee580d04cade99ff025ab3b4d&pid=1-s2.0-S1470160X1200132X-main.pdf.

Gallery Yopriceville (URL). http://driverlayer.com/showimg?v=g&img=http://gallery.yopriceville.com/downloadfullsize/send/3187&org=http://gallery.yopriceville.com/Free-Clipart-Pictures/Trees-PNG-Clipart/Fou_Seasons_Trees_PNG_Clipart, (17.07.2015).

Gallopín, Gilberto Carlos (1997): Indicators and Their Use: Information for Decision-making. Part One - Introduction. In: Moldan, Bedřich; Billharz, Suzanne (Eds.): Sustainability indicators. A report on the project on indicators of sustainable development. Chichester [England], New York: Wiley (SCOPE, 58), PDF pp 2-16. URL: http://stoa.usp.br/soniacoutinho/files/-1/19796/texto_15.pdf.

Gaussin, M. et al. (2013): Assessing the environmental footprint of manufactured products: A survey of current literature. In: International Journal of Production Economics, Vol. 146 Iss. 2, pp. 515–523. DOI: 10.1016/j.ijpe.2011.12.002. URL: http://www.sciencedirect.com/science/article/pii/S0925527311004981/pdfft?md5=215b2f126055d71bdc9698f8cd3a5d3c&pid=1-s2.0-S0925527311004981-main.pdf.

Gladen, Werner (2014): Performance Measurement. Controlling mit Kennzahlen. 6., überarb. Aufl. 2014. Wiesbaden: Springer Gabler (SpringerLink : Bücher). URL: http://link.springer.com/content/pdf/10.1007%2F978-3-658-05138-9.pdf.

Goldmann, Bernhard; Rauberger, Rainer and Seidel, Eberhard (2001): Umweltkennzahlen zur Unterstützung betrieblicher Entscheidungen. 7.2. In: Bundesumweltministerium, Umweltbundesamt (Eds.): Handbuch Umweltcontrolling. 2nd ed. München: Vahlen, pp. 597–623.

Govindan, Kannan et al. (2014): Impact of supply chain management practices on sustainability. In: Journal of Cleaner Production, Vol. 85, pp. 212–225. DOI: 10.1016/j.jclepro.2014.05.068. URL: http://www.sciencedirect.com/science/article/pii/S0959652614005460/pdfft?md5=d9498a4a4bcb47d20edd9f26356ee4a4&pid=1-s2.0-S0959652614005460-main.pdf.

Govindan, Kannan; Khodaverdi, Roohollah and Jafarian, Ahmad (2013): A fuzzy multi criteria approach for measuring sustainability performance of a supplier based on triple bottom line approach. In: Journal of Cleaner Production, Vol. 47, pp. 345–354. DOI: 10.1016/j.jclepro.2012.04.014. URL: http://www.sciencedirect.com/science/article/pii/S0959652612002016/pdfft?md5=93876bb6c1e9702dcf13989e977da594&pid=1-s2.0-S0959652612002016-main.pdf.

GRI (2000-2011): IP. Indicator Protocols Set Environment (EN). Version 3.1. Global Reporting Initiative (GRI). WRI; WBCSD. URL: https://www.globalreporting.org/resourcelibrary/G3.1-Environment-Indicator-Protocols.pdf.

GRI (2013): Implementation Manual. G4 Sustainability Reporting Guidelines. Global Reporting Inititative (GRI). URL: https://www.globalreporting.org/resourcelibrary/GRIG4-Part2-Implementation-Manual.pdf.

Groupe Casino (URL): Encouraging consumption that respects the environment. http://www.groupe-casino.fr/en/responsible-retailer/encouraging-consumption-that-respects-the-environment/ (14.06.2015).

Guardian (URL): Wheat prices climb as heat takes toll on crops. Guardian Media Limited. http://www.classifieds.guardian.co.tt/business/2012-06-28/wheat-prices-climb-heat-takes-toll-crops (10.07.2015).

Guide, V. Daniel R. and van Wassenhove, Luk N. (2001): Managing Product Returns for Remanufacturing. In: Production and Operations Management, Vol. 10 Iss. 2, PDF pp. 1-19. DOI: 10.1111/j.1937-5956.2001.tb00075.x. URL: http://flora.insead.edu/fichiersti_wp/inseadwp2000/2000-35.pdf.

Günther, Edeltraud and Sturm, Anke (2002): Environmental Performance Measurement. In: Bennett, Martin, Bouma, Jan Jaap; Wolters, Teun (Eds.): Environmental Management Accounting: Informational and Institutional Developments. Dordrecht [etc.]: Kluwer Academic Publishers (Eco-Efficiency in Industry and Science, 9), pp. 223–229.

Haanaes, Knut et al. (URL): Making Sustainability Profitable. Financial Management. Harvard Business Review. https://hbr.org/2013/03/making-sustainability-profitable (07.07.2014).

Hagelaar, Geoffrey J.L.F and van der Vorst, Jack G.A.J (2002): Environmental supply chain management: using life cycle assessment to structure supply chains. In: The International Food and Agribusiness Management Review, Vol. 4 Iss. 4, pp. 399–412. DOI: 10.1016/S1096-7508(02)00068-X. URL: http://elogistics.lhu.edu.tw/may/course/logistics/914b/pdf/35-36.pdf.

Hahn, Tobias and Scheermesser, Mandy (2006): Approaches to corporate sustainability among German companies. In: Corp. Soc. Responsib. Environ. Mgmt (Corporate Social Responsibility and Environmental Management), Vol. 13 Iss. 3, PDF pp. 2-17. DOI: 10.1002/csr.100. URL: http://www.researchgate.net/profile/Tobias_Hahn5/publication/227538194_Approaches_to_corporate_sustainability_among_German_companies/links/0deec5212398bb1832000000.pdf.

Hallay, Hendric and Pfriem, Reinhard (1992): Öko-Controlling. Umweltschutz in mittelständischen Unternehmen. Frankfurt a.M. [etc.]: Campus Verl.

Hallward, John (2008): “Make Measurable What Is Not So”: Consumer Mix Modeling for the Evolving Media World. In: J. Adv. Res. (Journal of Advertising Research), Vol. 48 Iss. 3, pp. 339–351. DOI: 10.2501/S0021849908080392. URL: http://web.a.ebscohost.com/ehost/viewarticle?data=dGJyMPPp44rp2%2fdV0%2bnjisfk5Ie45PFIr6m2T7Kk63nn5Kx95uXxjO62tUewpq1KrqezOLewsk64qq84v8OkjPDX7Ivf2fKB7eTnfLujr0%2b0r69Ot6uySKTi34bls%2bOGpNrgVeDr5j7y1%2bVVv8SkeeyzsUyzqLZLrqykfu3o63nys%2b585LzzhOrK45Dy&hid=4204.

Handfield, Robert et al. (2002): Applying environmental criteria to supplier assessment: A study in the application of the Analytical Hierarchy Process. In: European Journal of Operational Research, Vol. 141 Iss. 1, pp. 70–87. DOI: 10.1016/S0377-2217(01)00261-2. URL: http://www.sciencedirect.com/science/article/pii/S0377221701002612/pdfft?md5=3c89337816b7b59bff22c08db8e08500&pid=1-s2.0-S0377221701002612-main.pdf.

Hang, Song and Chunguang, Zhao (2015): Does environmental management improve enterprise's value? – An empirical research based on Chinese listed companies. In: Ecological Indicators, Vol. 51, pp. 191–196. DOI: 10.1016/j.ecolind.2014.08.020. URL: http://www.sciencedirect.com/science/article/pii/S1470160X14003756/pdfft?md5=f03e27e175798fd0dab0095897f4e8fe&pid=1-s2.0-S1470160X14003756-main.pdf.

Hansgrohe SE (E-Mail): E-Mail conversation with AKa, Production and Sustainability Controlling, Hansgrohe SE, Schiltach, Germany, 06.07.2015.

Harel, Tamar (2013): The Journey Of A Lifetime. Interface. …describes the role of The Natural Step Framework in Interface’s journey towards Mission Zero. The Natural Step (Ed.). URL: http://www.naturalstep.ca/sites/default/files/case_study_interface.pdf.

Harrison, Jeffrey S. and Wicks, Andrew C. (2013): Stakeholder Theory, Value, and Firm Performance. In: Business Ethics Quarterly, Vol. 23 Iss. 1, pp. 97–124. DOI: 10.5840/beq20132314. URL: http://centres.insead.edu/social-innovation/what-we-do/documents/stakehodertheory.pdf.

Harvey, Fiona (URL): England faces wildlife tragedy as worst drought in 30 years hits habitats. The Guardian. http://www.theguardian.com/environment/2012/mar/19/england-wildlife-drought (22.04.2015).

Hashemi, Seyed Hamid; Karimi, Amir and Tavana, Madjid (2015): An integrated green supplier selection approach with analytic network process and improved Grey relational analysis. In: International Journal of Production Economics, Vol. 159, pp. 178–191. DOI: 10.1016/j.ijpe.2014.09.027. URL: http://www.sciencedirect.com/science/article/pii/S0925527314003053/pdfft?md5=7d35be6c90a0cac22175ebde2d5c504c&pid=1-s2.0-S0925527314003053-main.pdf.

Haubach, Christian (2013): Umweltmanagement in globalen Wertschöpfungsketten. Eine Analyse am Beispiel der betrieblichen Treibhausgasbilanzierung. Wiesbaden: Springer Gabler. URL: http://link.springer.com/content/pdf/10.1007%2F978-3-658-02487-1.pdf.

Hauff, Michael von and Kleine, Alexandro (2009): Nachhaltige Entwicklung. Grundlagen und Umsetzung. München: Oldenbourg.

Hawken, Paul (1996): Kollaps oder Kreislaufwirtschaft. Wachstum nach dem Vorbild der Natur. 1st ed. Berlin: Siedler.

Hay, Buddy (E-Mail): E-Mail conversation with Buddy Hay, Assistant Vice President, Sustainable Strategies, Interface, Inc., Atlanta, Georgia USA, 24.06.2015. E-Mail Attachment: Two PDF documents: Interface EcoMetrics 2014 and Interface 2014 Sustainability Highlights.

Heinberg, Richard (2004): The Party's Over. Das Ende der Ölvorräte und die Zukunft der industrialisierten Welt. 1st ed. München: Riemann (One earth spirit).

Hermann, B.G; Kroeze, C. and Jawjit, W. (2007): Assessing environmental performance by combining life cycle assessment, multi-criteria analysis and environmental performance indicators. In: Journal of Cleaner Production, Vol. 15 Iss. 18, pp. 1787–1796. DOI: 10.1016/j.jclepro.2006.04.004. URL: http://www.sciencedirect.com/science/article/pii/S0959652606001569/pdfft?md5=2a6a9577dc27dc66e6a31f1ad465f076&pid=1-s2.0-S0959652606001569-main.pdf.

Herva, Marta et al. (2011): Review of corporate environmental indicators. In: Journal of Cleaner Production, Vol. 19 Iss. 15, pp. 1687–1699. DOI: 10.1016/j.jclepro.2011.05.019. URL: http://www.sciencedirect.com/science/article/pii/S0959652611001892/pdfft?md5=3beb54ac9c5fe361a7a38f7f579c4570&pid=1-s2.0-S0959652611001892-main.pdf.

Hervani, Aref A.; Helms, Marilyn M. and Sarkis, Joseph (2005): Performance measurement for green supply chain management. In: Benchmarking, Vol. 12 Iss. 4, pp. 330–353. DOI: 10.1108/14635770510609015. URL: http://www.emeraldinsight.com/doi/pdfplus/10.1108/14635770510609015.

Herzig, Christian et al. (2012): Environmental Management Accounting. Case studies in South-East Asian companies. London, New York: Routledge.

Hesse, Axel (2010): SD-KPI standard 2010 - 2015. Sustainable development key performance indicators (SD-KPIs); minimum reporting standard for relevant sustainability information in annual reports; management commentaries of 68 industries. Version 1.2. Münster, Am Krug 30: A. Hesse. URL: http://www.sd-m.de/files/SD-KPI_Standard_2010-2015_V2d.pdf.

Hockerts, Kai (2001): Corporate Sustainability Management. Towards Controlling Corporate Ecological And Social Sustainability. In: Proceedings of Greening of Industry Network Conference. 21-24 January 2001 Bangkok, PDF pp. 1-21. URL: https://gin.confex.com/gin/archives/2001/papers/25.pdf.

Hoekstra, A. Y. and Chapagain, A. K. (2006): Water footprints of nations: Water use by people as a function of their consumption pattern. In: Water Resour Manage (Water Resources Management), Vol. 21 Iss. 1, pp. 35–48. DOI: 10.1007/s11269-006-9039-x. URL: http://link.springer.com/content/pdf/10.1007%2Fs11269-006-9039-x.pdf.

Hoffmann, Tim (2011): Unternehmerische Nachhaltigkeitsberichterstattung. Eine Analyse des GRI G3.1-Berichtsrahmens. Lohmar: Eul (Rechnungslegung und Wirtschaftsprüfung, 30).

Hopfenbeck, Waldemar and Jasch, Christine (1993): Öko-Controlling. Umdenken zahlt sich aus! Audits, Umweltberichte und Ökobilanzen als betriebliche Führungsinstrumente. Landsberg/Lech: Verl. Moderne Industrie.

Hörisch, Jacob; Windolph, Sarah Elena and Schaltegger, Stefan (2014): Corporate Sustainability Management in Large German Companies. Chapter 7. In: Schaltegger, Stefan; Windolph, Sarah Elena; Harms, Dorli; Hörisch, Jacob (Eds.): Corporate Sustainability in International Comparison. State of Practice, Opportunities and Challenges, Vol. 31. Switzerland: Springer (Eco-Efficiency in Industry and Science, 31), pp. 93–104.

Horváth, Péter (2009): Controlling. 11th ed. München: Vahlen (Vahlens Handbücher der Wirtschafts- und Sozialwissenschaften).

Horváth, Péter; Isensee, Johannes and Michel, Uwe (2012): "Green Controlling" - Bedarf einer Integration von ökologischen Aspekten in das Controlling. In: Tschandl, Martin; Posch, Alfred (Eds.): Integriertes Umweltcontrolling. Von der Stoffstromanalyse zum Bewertungs- und Informationssystem. 2nd ed. Wiesbaden: Gabler Verlag / Springer Fachmedien Wiesbaden GmbH, Wiesbaden (SpringerLink : Bücher), pp. 41–50. URL: http://link.springer.com/content/pdf/10.1007%2F978-3-8349-6844-9.pdf.

Hsu, Chia-Wei and Hu, Allen H. (2009): Applying hazardous substance management to supplier selection using analytic network process. In: Journal of Cleaner Production, Vol. 17 Iss. 2, pp. 255–264. DOI: 10.1016/j.jclepro.2008.05.004. URL: http://www.coeng.ntut.edu.tw/ezfiles/0/academic/22/academic_50753_8415397_56406.pdf.

Huang, Samuel H. and Keskar, Harshal (2007): Comprehensive and configurable metrics for supplier selection. In: International Journal of Production Economics, Vol. 105 Iss. 2, pp. 510–523. DOI: 10.1016/j.ijpe.2006.04.020. URL: http://www.sciencedirect.com/science/article/pii/S0925527306001241/pdfft?md5=02f6cf8444b2d1649bbd1a251077447d&pid=1-s2.0-S0925527306001241-main.pdf.

Hubert, Boris (2015): Controlling-Konzeptionen. Ein schneller Einstieg in Theorie und Praxis: Springer Gabler (Essentials). URL: http://link.springer.com/content/pdf/10.1007%2F978-3-658-07565-1.pdf.

Humphreys, Paul; McIvor, Ronan and Chan, Felix (2003): Using case-based reasoning to evaluate supplier environmental management performance. In: Expert Systems with Applications, Vol. 25 Iss. (2), pp. 141–153. DOI: 10.1016/S0957-4174(03)00042-3. URL: http://www.sciencedirect.com/science/article/pii/S0957417403000423/pdfft?md5=9901fd209b98ed474b418d680aeaa361&pid=1-s2.0-S0957417403000423-main.pdf.

Hunsberger, Carol and Evans, Tom P. (2012): Land. Chapter 3. In: UNEP (Ed.): Global Environment Outlook GEO 5. Environment for the future we want. Nairobi, Kenya: United Nations Environment Programme, pp. 65–96. URL: http://www.unep.org/geo/pdfs/geo5/GEO5_report_full_en.pdf.

Hutchins, Margot J. and Sutherland, John W. (2008): An exploration of measures of social sustainability and their application to supply chain decisions. In: Journal of Cleaner Production, Vol. 16 Iss. 15, pp. 1688–1698. DOI: 10.1016/j.jclepro.2008.06.001. URL: http://www.researchgate.net/profile/John_Sutherland5/publication/242211332_An_exploration_of_measures_of_social_sustainability_and_their_application_to_supply_chain_decisions/links/00b7d528eca9e3b34f000000.pdf.

ICAEW (URL): Environmental key performance indicators (KPI). A glossary of environmental key performance indicators (KPIs). http://www.icaew.com/en/technical/business-performance-management-community/data-and-kpis/kpi-library/environmental-kpis (11.06.2015).

IFAC (2005): Environmental Management Accounting. International Guidance Document. New York, N.Y: International Federation of Accountants (IFAC). URL: http://www.ioew.at/ioew/download/IFAC_Guidance_Env_Man_Acc_0508.pdf.

Jacobson, Robert (1987): The Validity of ROI as a Measure of Business Performance. In: The American Economic Review, Vol. 77 Iss. 3, pp. 470–478. URL: http://web.b.ebscohost.com/ehost/viewarticle?data=dGJyMPPp44rp2%2fdV0%2bnjisfk5Ie46a9JrquzT7Ck63nn5Kx95uXxjL6srUmxpbBIr6ueTbiosFKzpp5Zy5zyit%2fk8Xnh6ueH7N%2fiVauorkuvrLZPt6mvPurX7H%2b72%2bw%2b4ti7evLepIzf3btZzJzfhruqs0izqLFQpNztiuvX8lXu2uRe8%2bLqbOPu8gAA&hid=102.

Jäger, Jill and Patel, Neeyati (2012): An Earth System Perspective. Chapter 7. In: UNEP (Ed.): Global Environment Outlook GEO 5. Environment for the future we want. Nairobi, Kenya: United Nations Environment Programme, pp. 193–214. URL: http://www.unep.org/geo/pdfs/geo5/GEO5_report_full_en.pdf.

Jasch, Christine (2000): Environmental performance evaluation and indicators. In: Journal of Cleaner Production, Vol. 8 Iss. 1, pp. 79–88. DOI: 10.1016/S0959-6526(99)00235-8. URL: http://www.sciencedirect.com/science/article/pii/S0959652699002358/pdfft?md5=59a94be911b42385e2b7dff17b0be542&pid=1-s2.0-S0959652699002358-main.pdf.

Jasch, Christine (2002): Environmental Management Accounting Metrics: Procedures and Principles. In: Bennett, Martin; Bouma, Jan Jaap; Wolters, Teun (Eds.): Environmental Management Accounting: Informational and Institutional Developments. Dordrecht [etc.]: Kluwer Academic Publishers (Eco-Efficiency in Industry and Science, 9), pp. 37–50.

Jasch, Christine (2009): Environmental and material flow cost accounting. Principles and procedures. [New York], [Vienna, Austria], [Holland]: Springer; IÖW; EMAN (Eco-Efficiency in Industry and Science, 25). URL: http://link.springer.com/content/pdf/10.1007%2F978-1-4020-9028-8.pdf.

Jennen, Jürgen and Zimmermann, Klaus (1976): Ein Ansatz zur Konstruktion eines Umweltgesamtindikators. Entwicklungsmöglichkeiten und Grenzen. Berlin (Papers aus dem Internationales Institut für Umwelt und Gesellschaft des Wissenschaftszentrums Berlin, I/76-8).

Kaplan, Robert S. and Norton, David P. (1996): The Balanced Scorecard. Translating Strategy Into Action. Boston, Massachusetts: Harvard Business School Press.

Kates, Robert W. (2015): Sustainability and Sustainability Science. In: International Encyclopedia of the Social & Behavioral Sciences: Elsevier, pp. 801–806. DOI: 10.1016/B978-0-08-097086-8.91028-6. URL: http://www.sciencedirect.com/science/article/pii/B9780080970868910286/pdfft?md5=703bfe3d14190cdcd61f45772a506b5e&pid=3-s2.0-B9780080970868910286-main.pdf.

Khan, Zaheer and Nicholson, John D. (2014): An investigation of the cross-border supplier development process: Problems and implications in an emerging economy. In: International Business Review, Vol. 23 Iss. 6, pp. 1212–1222. DOI: 10.1016/j.ibusrev.2014.05.001. URL: http://www.sciencedirect.com/science/article/pii/S0969593114000547/pdfft?md5=a11cd697857bf130720a904027945417&pid=1-s2.0-S0969593114000547-main.pdf.

Kinderyte, Loreta; Ciegis, Remigijus and Staniskis, Jurgis Kazimieras (2010): Assessment Of Enterprise Performance For Efficient Sustainability Management. In: Transformations in Business and Economics, Vol. 9 Iss. 3, pp. 104–118. URL: http://web.a.ebscohost.com/ehost/viewarticle?data=dGJyMPPp44rp2%2fdV0%2bnjisfk5Ie46a9JrquzT7Ck63nn5Kx95uXxjL6srUqtqK5Js5a3Uq%2bsuEyvlr9lpOrweezp33vy3%2b2G59q7RbOqsE6zqbVQt5zqeezdu33snOJ6u9vtgKTq33%2b7t8w%2b3%2bS7Sa%2brskqzqqR%2b7ejrefKz7nzkvPOE6srjkPIA&hid=4104.

KPMG International (2013): The KPMG Survey of Corporate Responsibility Reporting 2013. kpmg.com/sustainability. URL: http://www.kpmg.com/Global/en/IssuesAndInsights/ArticlesPublications/corporate-responsibility/Documents/corporate-responsibility-reporting-survey-2013-v2.pdf.

Krajnc, Damjan and Glavic, Peter (2003): Indicators of sustainable production. In: Clean Technologies and Environmental Policy, Vol. 5 Iss. 3-4, pp. 279–288. DOI: 10.1007/s10098-003-0221-z. URL: http://link.springer.com/content/pdf/10.1007%2Fs10098-003-0221-z.pdf.

Kühnapfel, Jörg B. (2014): Balanced Scorecards im Vertrieb. Wiesbaden: Springer Gabler (Essentials). URL: http://link.springer.com/content/pdf/10.1007%2F978-3-658-05496-0.pdf.

Kumar, Amit; Jain, Vipul and Kumar, Sameer (2014): A comprehensive environment friendly approach for supplier selection. In: Omega, Vol. 42 Iss. 1, pp. 109–123. DOI: 10.1016/j.omega.2013.04.003. URL: http://www.sciencedirect.com/science/article/pii/S0305048313000522/pdfft?md5=fba3a0549fac14126486254bd3d5a67d&pid=1-s2.0-S0305048313000522-main.pdf.

Kuylenstierna, Johan C.I and Ajero, May Antoniette (2012): Atmosphere. Chapter 2. In: UNEP (Ed.): Global Environment Outlook GEO 5. Environment for the future we want. Nairobi, Kenya: United Nations Environment Programme, pp. 31–64. URL: http://www.unep.org/geo/pdfs/geo5/GEO5_report_full_en.pdf.

Lahme, Georg and Klenk, Volker (2014): Klenk & Hoursch: Telling the Backstory - Transparenz in globalen Wertschöpfungsketten. In: D'heur, Michael (Ed.): CSR und Value Chain Management. Profitables Wachstum durch nachhaltig gemeinsame Wertschöpfung. Berlin, Heidelberg: Springer Gabler (SpringerLink : Bücher), pp. 155–171. URL: http://link.springer.com/content/pdf/10.1007/978-3-642-39889-6.pdf.

Lee, Amy H.I et al. (2009): A green supplier selection model for high-tech industry. In: Expert Systems with Applications, Vol. 36 Iss. 4, pp. 7917–7927. DOI: 10.1016/j.eswa.2008.11.052. URL: http://www.sciencedirect.com/science/article/pii/S095741740800821X/pdfft?md5=add487b0d2148f5a2bedb0e87abb18a9&pid=1-s2.0-S095741740800821X-main.pdf.

Levy, D. L.; Szejnwald Brown, H. and Jong, M. de (2010): The Contested Politics of Corporate Governance: The Case of the Global Reporting Initiative. In: Business & Society, Vol. 49 Iss. 1, pp. 88–115. DOI: 10.1177/0007650309345420. URL: http://bas.sagepub.com/content/49/1/88.full.pdf.

Linke, Barbara S. et al. (2013): Sustainability indicators for discrete manufacturing processes applied to grinding technology. In: Journal of Manufacturing Systems, Vol. 32 Iss. 4, pp. 556–563. DOI: 10.1016/j.jmsy.2013.05.005. URL: http://www.sciencedirect.com/science/article/pii/S0278612513000708/pdfft?md5=66cac86f8a9a53e85a349b31a400b415&pid=1-s2.0-S0278612513000708-main.pdf.

Linnenluecke, Martina K. and Griffiths, Andrew (2012): Firms and sustainability: Mapping the intellectual origins and structure of the corporate sustainability field. In: Global Environmental Change, Vol. 23 Iss. 1, pp. 382–391. DOI: 10.1016/j.gloenvcha.2012.07.007. URL:http://www.sciencedirect.com/science/article/pii/S095937801200091X/pdfft?md5=9e80a12bc1bcf46d5ef9f8fbd9a982c6&pid=1-s2.0-S095937801200091X-main.pdf.

Lisoski, Ed (2003): if you can't measure it you can't manage it. In: Supervision, Vol. 64 Iss. 64, pp. 14–17. URL: http://web.a.ebscohost.com/ehost/viewarticle?data=dGJyMPPp44rp2%2fdV0%2bnjisfk5Ie45PFIr6m2T7Kk63nn5Kx95uXxjO62tUewpq1KrqezOLewsk64qq84v8OkjPDX7Ivf2fKB7eTnfLujr0%2b0r69Ot6uySKTi34bls%2bOGpNrgVeDr5j7y1%2bVVv8Skeeyztk%2b1q65OsJzkh%2fDj34y75uJ%2bxOvqhNLb9owA&hid=4204.

Liverman, Diana M. et al. (1988): Global Sustainability: Toward Measurement. In: Environmental Management, Vol. 12 Iss. 2, pp. 133–143. DOI: 10.1007/BF01873382. URL: http://link.springer.com/content/pdf/10.1007/BF01873382.pdf.

Lourenço, Isabel C. and Branco, Manuel Castelo (2013): Determinants of corporate sustainability performance in emerging markets: the Brazilian case. In: Journal of Cleaner Production, Vol. 57, pp. 134–141. DOI: 10.1016/j.jclepro.2013.06.013. URL: http://www.sciencedirect.com/science/article/pii/S0959652613003946/pdfft?md5=06484d69ae3f14673cd4f9fe02d0c487&pid=1-s2.0-S0959652613003946-main.pdf.

Lu, Louis Y.Y; Wu, C.H and Kuo, T.-C (2007): Environmental principles applicable to green supplier evaluation by using multi-objective decision analysis. In: International Journal of Production Research, Vol. 45 Iss. 18-19, pp. 4317–4331. DOI: 10.1080/00207540701472694. URL: http://web.a.ebscohost.com/ehost/viewarticle?data=dGJyMPPp44rp2%2fdV0%2bnjisfk5Ie45PFIr6m2T7Kk63nn5Kx95uXxjO62tUevrK1KrqezOLCwsU%2b4prE4zsOkjPDX7Ivf2fKB7eTnfLuntUiwrbdPsqe2PurX7H%2b72%2bw%2b4ti7evPepIzf3btZzJzfhruotEyyra9Rt5zkh%2fDj34y75uJ%2bxOvqhNLb9owA&hid=4209.

Lundin, Margareta and Morrison, Gregory M. (2002): A life cycle assessment based procedure for development of environmental sustainability indicators for urban water systems. In: Urban Water, Vol. 4 Iss. 2, pp. 145–152. DOI: 10.1016/S1462-0758(02)00015-8. URL: http://www.sciencedirect.com/science/article/pii/S1462075802000158/pdfft?md5=93fae167e5758bba7f9a49ce80ab1372&pid=1-s2.0-S1462075802000158-main.pdf.

Mani, V.; Agrawal, Rajat and Sharma, Vinay (2014): Supplier selection using social sustainability: AHP based approach in India. In: International Strategic Management Review, Vol. 2 Iss. 2, pp. 98–112. DOI: 10.1016/j.ism.2014.10.003. URL: http://www.sciencedirect.com/science/article/pii/S2306774814000209/pdfft?md5=d5541cb5606288fac4986d893dc3ef8e&pid=1-s2.0-S2306774814000209-main.pdf.

Maria, Sperdea Natalita; Mirela, Mazilu and Cristina, Marinescu Roxana (2010): The Management Of The Optimal Conditions Of Storage - Transport - Trading Of The Food Products. In: Annals of Faculty of Economics, Vol. 1 Iss. 2, pp. 1013–1019. URL: http://anale.steconomiceuoradea.ro/volume/2010/n2/161.pdf.

Martin, Andrew (URL): How green is my orange? The New York Times. http://www.nytimes.com/2009/01/22/business/worldbusiness/22iht-22pepsi.19583527.html (14.06.2015).

McIntyre, Kirstie et al. (1998): Environmental performance indicators for integrated supply chains: the case of Xerox Ltd. In: Supp Chain Mnagmnt (Supply Chain Management: An International Journal), Vol. 3 Iss. 3, pp. 149–156. DOI: 10.1108/13598549810230877. URL: http://www.emeraldinsight.com/doi/pdfplus/10.1108/13598549810230877.

Meadows, Donella H. et al. (1972): The Limits to Growth. A Report for the Club of Rome's Project on the Predicament of Mankind. New York: Universe Books (Potomac Associates books).

Mentzer, John T. et al. (2001): Defining Supply Chain Mangement. In: Journal of Business Logistics, Vol. 22 Iss. 2, pp. 1–25. DOI: 10.1002/ j.2158-1592.2001.tb00001.x. URL: http://biblioteca.fundacionicbc.edu.ar/ images/e/e4/Conexion_y_logistica_2.pdf.

Michael, Florianna L.; Noor, Zainura Zainon and Figueroa, Maria J. (2014): Review of urban sustainability indicators assessment – Case study between Asian countries. In: Habitat International, Vol. 44, pp. 491–500. DOI: 10.1016/j.habitatint.2014.09.006. URL: http://www.sciencedirect.com/science/ article/pii/S0197397514001350/pdfft?md5=ee2d6dbcc1fe1e4c5063a30d2de98 ee7&pid=1-s2.0-S0197397514001350-main.pdf.

Michel, Uwe; Isensee, Johannes and Stehle, Alexander (2014): Sustainability Controlling: Planung, Steuerung und Kontrolle der Realisierung der Nachhaltigkeitsstrategie. In: Schulz, Thomas; Bergius, Susanne (Eds.): CSR und Finance. Beitrag und Rolle des CFO für eine Nachhaltige Unternehmensführung. Berlin, Heidelberg: Springer Gabler (Management-Reihe Corporate Social Responsibility), pp. 97–111. URL: http://link.springer.com/content/pdf/10.1007/978-3-642-54882-6.pdf.

Moneim, Ahmed Farouk Abdul; Galal, Noha M. and Shakwy, Mohamed El (2013): Sustainable Manufacturing Indicators. In: Global Climate Change, PDF pp. 1-12. URL: http://gccbs2013.aast.edu/newgcc/images/pdf/ sustainable%20manufacturing%20indicators.pdf.

Müller, Armin (2011): Nachhaltigkeits-Controlling. 1st ed. Berlin: Uni-Edition (Markt- und werteorientierte Unternehmensführung, 6).

National Energy Foundation (URL): Simple Carbon Calculator. CO2 Calculator. National Energy Foundation 2001-15. http://www.nef.org.uk/greencompany/ co2calculator.htm (20.07.2015).

Neely, Andy; Micheli, Pietro and Martinez, Veronica (2006): Acting on Information. Performance management for the public sector. London: Advanced Institute of Management Research. URL: https://dspace.lib.cranfield.ac.uk/bitstream/ 1826/4254/1/Acting_on_information.pdf.

Nieters, Anne; Drosdowski, Thomas and Lehr, Ulrike (2015): Do extreme weather events damage the German economy? Gesellschaft für Wirtschaftliche Strukturforschung mbH. Osnabrück (gws Discussion Paper 2015/2). URL: http://www.gws-os.com/discussionpapers/gws-paper15-2.pdf.

Nikbakhsh, Ehsan (2009): Green Supply Chain Management. Chapter 9. In: Farahani, Reza Zanjirani; Asgari, Nasrin; Davarzani, Hoda (Eds.): Supply Chain and Logistics in National, International and Governmental Environment. Heidelberg: Physica-Verlag HD, pp. 195–220. URL: http://link.springer.com/content/pdf/10.1007%2F978-3-7908-2156-7.pdf.

Noci, Giuliano (1997): Designing 'green' vendor rating systems for the assessment of a supplier's environmental performance. In: European Journal of Purchasing & Supply Management, Vol. 3 Iss. 2, pp. 103–114. DOI: 10.1016/S0969-7012(96)00021-4. URL: http://www.sciencedirect.com/science/article/pii/S0969701296000214/pdf?md5=227fc775c9463b29c37117e2aa201c65&pid=1-s2.0-S0969701296000214-main.pdf.

Norman, Wayne and MacDonald, Chris (2004): Getting to the Bottom of "Triple Bottom Line". In: Business Ethics Quarterly, Vol. 14 Iss. 2, pp. 243–262. URL: http://web.a.ebscohost.com/ehost/viewarticle?data=dGJyMPPp44rp2%2fdV0%2bnjisfk5Ie45PFIr6m2T7Kk63nn5Kx95uXxjO62tUevra1KrqezOLOwsUu4p7c4v8OkjPDX7Ivf2fKB7eTnfLuptkuvqrVIsaqkhN%2fk5VXj5KR84LPgjeac8nnls79mpNfsVa%2botUu1q7NQpNztiuvX8lXu2uRe8%2bLqbOPu8gAA&hid=4209.

NZBCSD (2002): The challenge of greenhouse gas emissions. The "why" and "how" of accounting and reporting for GHG emissions. An Industry Guide. New Zealand Business Council for Sustainable Development (NZBCSD). Auckland. URL: http://www.sbc.org.nz/__data/assets/pdf_file/0019/54901/Climate_Change_Guide.pdf.

OECD (2011): Sustainable Manufacturing Toolkit. Seven Steps to Environmental Excellence. Start-Up Guide. URL: http://www.oecd.org/innovation/green/toolkit/48704993.pdf.

OECD (URL1): Glossary Of Statistical Terms. Carbon Dioxide Equivalent. https://stats.oecd.org/glossary/detail.asp?ID=285 (05.06.2015).

OECD (URL2): OECD sustainable manufacturing indicators. Explaining Indicators. In order to see the respective indicators' formulas and explanations click on the stated link titles. http://www.oecd.org/innovation/green/toolkit/oecdsustainablemanufacturingindicators.htm (06.06.2015).

Olsthoorn, Xander et al. (2001): Environmental indicators for business: a review of the literature and standardisation methods. In: Journal of Cleaner Production, Vol. 9 Iss. 5, pp. 453–463. DOI: 10.1016/S0959-6526(01)00005-1. URL: http://www.sciencedirect.com/science/article/pii/S0959652601000051/pdfft?md5=497ea2b08d403a0c88c852b188a6845b&pid=1-s2.0-S0959652601000051-main.pdf.

Øvergaard, Sara (2008): Issue Paper: Definition of primary and secondary energy. Statistics Norway, Oslo. URL: http://millenniumindicators.un.org/unsd/envAccounting/londongroup/meeting13/LG13_12a.pdf.

Öztürk, Burcu Avcı and Özçelik, Funda (2014): Sustainable Supplier Selection with A Fuzzy Multi-Criteria Decision Making Method Based on Triple Bottom Line. In: Business & Economics Research Journal, Vol. 5 Iss. 3, pp. 129–147. URL: http://www.berjournal.com/wp-content/plugins/downloads-manager/upload/BERJ5(3)14Article8pp.129-147.pdf.

P&G (2014): P&G Supply Chain Environmental Sustainability Scorecard. Supply Chain Environmental Sustainability Scorecard. P&G Suppliers. Procter & Gamble (P&G). URL: http://www.pgsupplier.com/en/current-suppliers/environmental-sustainability-scorecard.shtml (25/05/2015). Excel File: http://www.pgsupplier.com/en/_downloads/SCESS_2014.xlsx, Excel sheet 1: Purpose, Excel sheet 2: Instructions, Excel sheet 3: Scorecard, Excel sheet 4: Rating Criteria.

P&G (URL1): Improving sustainability end to end. Helping Suppliers Conserve. Procter & Gamble (P&G). http://www.pg.com/en_US/sustainability/environmental_sustainability/conservation_resources/helping_suppliers.shtml (27.05.2015).

P&G (URL2): P&G Shares Data, Results and Analysis Tool from Supplier Scorecard. Scorecard Fosters Collaboration, Innovation; Savings from Sustainability Efforts Near $1BN. P&G Corporate Newsroom. NYSE: PG (Ed.). Procter & Gamble (P&G). Cincinnati. http://news.pg.com/press-release/pg-corporate-announcements/pg-shares-data-results-and-analysis-tool-supplier-scorecard (24.05.2015).

Park, Penny (2015): The Emergence of Infectious Diseases. In: UNEP (Ed.): UNEP Year Book 2014. Emerging Issues in Our Global Environment: United Nations Pubns, pp. 12–17. URL: http://www.unep.org/yearbook/2014/PDF/UNEP_YearBook_2014.pdf.

Peccoud, Jean and Alexandrov, Theodore P. (2014): If You Can't Measure It, You Can't Manage It. In: PLoS Comput Biol (PLoS Computational Biology), Vol. 10 Iss. 3, pp. 1-4. DOI: 10.1371/journal.pcbi.1003462. URL: http://web.a.ebscohost.com/ehost/viewarticle?data=dGJyMPPp44rp2%2fdV0%2bnjisfk5Ie45PFIr6m2T7Kk63nn5Kx95uXxjO62tUewp61KrqezOK%2bmuEy1sK9OnrfLPvLo34bx1%2bGM5%2bXsgeKzq1Gypq9Nt62xSaTi34bls%2bOGpNrgVd%2fm5j7y1%2bVVv8Skeeyzt02yqbJQt6mkfu3o63nys%2b585LzzhOrK45Dy&hid=4212.

Peña, Camilo Andrés and Salgado, Omar (2013): Sustainability indicators along the coffee value chain: A comparative study between Mexican & Colombian retail. In: 24th Annual Conference of the Production and Operations Management Society Denver Co. USA, PDF pp. 1-10. URL: http://www.pomsmeetings.org/confproceedings/043/FullPapers/FullPaper_files/043-0095.pdf.

Pérez-Calderón, E.; Milanés-Montero P. and Ortega-Rossell F. J. (2012): Environmental Performance and Firm Value: Evidence from Dow Jones Sustainability Index Europe. In: IJRE: International Journal of Environmental Research, Vol. 6 Iss. 4, pp. 1007–1014. URL: http://www.ijer.ir/pdf_571_1d7253e959774f486c5413eda321f847.html.

Porter, Michael E. (1985): Competitive Advantage. Creating and Sustaining Superior Performance. New York, London: Free Press; Collier Macmillan.

Probst, Hans-Jürgen (2006): Kennzahlen leicht gemacht. Richtig anwenden und interpretieren. Heidelberg: REDLINE Wirtschaft (Leicht gemacht).

Pulver, Simone (2007): Introduction: Developing-Country Firms as Agents of Environmental Sustainability? In: St Comp Int Dev (Studies in Comparative International Development), Vol. 42 Iss. 3-4, pp. 191–207. DOI: 10.1007/s12116-007-9011-7. URL: http://link.springer.com/content/pdf/10.1007%2Fs12116-007-9011-7.pdf.

PwC (2006): Guide to key performance indicators. Communicating the measures that matter. PricewaterhouseCoopers (Ed.). PwC. London. URL: http://www.pwc.com/gx/en/audit-services/corporate-reporting/assets/pdfs/UK_KPI_guide.pdf.

Rainey, David L. (2010): Sustainable Business Development. Inventing the Future through Strategy, Innovation, and Leadership. Pbk. reissue. Cambridge, New York: Cambridge University Press.

Rao, Purba Halady (2008): Greening The Supply Chain. A Guide For Asian Managers. New Delhi, Thousand Oaks, Calif: Response Books.

Rao, Purba (2002): Greening the supply chain: a new initiative in South East Asia. In: Int Jrnl of Op & Prod Mnagemnt (International Journal of Operations & Production Management), Vol. 22 Iss. 6, pp. 632–655. DOI: 10.1108/01443570210427668. URL: http://www.emeraldinsight.com/doi/pdfplus/10.1108/01443570210427668.

Rao, Purba H. (2014): Measuring Environmental Performance across a Green Supply Chain: A Managerial Overview of Environmental Indicators. In: VIKALPA, Vol. 39 Iss. 1, pp. 57–74. URL: http://www.vikalpa.com/pdf/articles/2014/vikalpa-39-1-57-74.pdf.

Rao, Purba and Holt, Diane (2005): Do green supply chains lead to competitiveness and economic performance? In: Int Jrnl of Op & Prod Mnagemnt (International Journal of Operations & Production Management), Vol. 25 Iss. 9, pp. 898–916. URL: http://www.emeraldinsight.com/doi/pdfplus/10.1108/01443570510613956.

Rao, Purba and Kondo, Mari (2010): A Study to Explore the Link between Green Purchasing Initiatives and Business Performance. In: Great Lakes Herald, Vol. 4 Iss. 2, PDF pp. 1-20. URL: http://www.greatlakes.edu.in/pdf/Herald/Vol4/Astudytoexplorethelinkbetweengreenpurchasinginitiativesandbusinessperformance.pdf.

Rao, Purba et al. (2009): A Metric for Corporate Environmental Indicators … for Small and Medium Enterprises in the Philippines. In: Bus. Strat. Env. (Business Strategy and the Environment), Vol. 8 Iss. 1, pp. 14–31. URL: http://onlinelibrary.wiley.com/doi/10.1002/bse.555/pdf.

Rappaport, Alfred (1986): Creating shareholder value. The new standard for business performance. New York, London: Free Press; Collier Macmillan.

Reece, James S. and Cool, William R. (1978): Measuring investment center performance. In: Havard Business Review, Vol. 56 Iss. 3, pp. 28-176 (10p). URL: http://web.a.ebscohost.com/ehost/viewarticle?data=dGJyMPPp44rp2%2fdV0%2bnjisfk5Ie46a9JrquzT7Ck63nn5Kx95uXxjL6srUmxpbBIr6ueTbirsFKwqp5Zy5zyit%2fk8Xnh6ueH7N%2fiVauutFGvrbNIsaakhN%2fk5VXj5KR84LPgjOac8nnls79mpNfsVbGutE%2bzrbE%2b5OXwhd%2fqu4ji3MSN6uLSffbq&hid=4106.

Reichmann, Thomas (2011): Controlling mit Kennzahlen. Die systemgestützte Controlling-Konzeption mit Analyse- und Reportinginstrumenten. 8th ed. München: Vahlen (Controlling Competence).

Renner, Michael (2002): Breaking the Link Between Resources and Repression. Chapter 7. In: Flavin, Christopher; Annan, Kofi A. (Eds.): State of the world 2002. A Worldwatch Institute report on progress toward a sustainable society. Special World Summit ed. New York, NY: Norton, pp. 149–172. URL: http://www.globalchange.umich.edu/gctext/Inquiries/Inquiries_by_Unit/Module%20Activities/State%20of%20the%20World/Resources%20and%20Repression%202002.pdf.

Riesner, Wilhelm and Sommer, Andreas (2010): Ausgewählte Aspekte der Umwelttechnik. Schwerpunkt Energie. In: Kramer, Matthias (Ed.): Integratives Umweltmanagement. Systemorientierte Zusammenhänge zwischen Politik, Recht, Management und Technik. 1st ed. Wiesbaden: Gabler, pp. 476–484. URL: http://link.springer.com/content/pdf/10.1007%2F978-3-8349-8602-3.pdf.

Robertson, Mark (2012): On track for Rio+20? How are global companies responding to sustainability? EIRIS Sustainability Report. EIRIS Ltd (Ed.). URL: http://www.eiris.org/files/research%20publications/EIRISGlobalSustainbailityReport2012.pdf.

Robins, Nick and Roberts, Sarah (1997): Reaping the benefits: Trade opportunities for developing-country producers from sustainable consumption and production. In: Greener Management International, Iss. 19, EBSCO HOST HTML Full Text, without page numbers. URL: http://web.a.ebscohost.com/ehost/detail/detail?sid=7af5bd55-b9a4-4091-869a-428a4de1b3ba%40sessionmgr4001&vid=26&hid=4212&bdata=JnNpdGU9ZWhvc3QtbGl2ZQ%3d%3d#db=bth&AN=117288.

Rogers, Dale S. and Tibben-Lembke, Ronald S. (1999): Going Backwards. Reverse Logistics Trends and Practices. Reverse Logistics Executive Council. URL: gio.uniovi.es/documentos/bel-li/rogers.pdf.

Rubik, Frieder (2001): Umweltzeichen und Produktvermarktung. 4.3. In: Bundesumweltministerium, Umweltbundesamt (Eds.): Handbuch Umweltcontrolling. 2nd ed. München: Vahlen, pp. 281-295.

Rudawska, Edyta; Renko, Sanda and Bilan, Yuriy (2013): Sustainable Development: Concept, Interest Groups, Benefits and Global Challenges. In: IJAR (International Journal of Academic Research), Vol. 5 Iss. 6, pp. 83–86. DOI: 10.7813/2075-4124.2013/5-6/B.15. URL: http://web.b.ebscohost.com/ehost/viewarticle?data=dGJyMPPp44rp2%2fdV0%2bnjisfk5Ie45PFIr6m2T7Kk63nn5Kx95uXxjO62tUevpq1KrqezOLSwrk64qbU4v8OkjPDX7Ivf2fKB7eTnfLujsUqwqrZLtKu2PurX7H%2b72%2bw%2b4ti7ee7epIzf3btZzJzfhruvs0%2bwrLZQsJzkh%2fDj34y75uJ%2bxOvqhNLb9owA&hid=118.

Russ, Hilary (URL): Update 2 - New York, New Jersey put $71 bln price tag on Sandy. Chicago Tribune. Reuters. http://articles.chicagotribune.com/2012-11-26/news/sns-rt-storm-sandycost-nyc-update-211e8mq4me-20121126_1_sandy-costs-hurricane-sandy-new-york-governor (22.04.2015).

Saen, Reza Farzipoor (2009): A new approach for selecting slightly non-homogeneous vendors. In: J of Advances in Mgmt Research (Journal of Advances in Management Research), Vol. 6 Iss. 2, pp. 144–153. DOI: 10.1108/09727980911007172. URL: http://www.emeraldinsight.com/doi/pdfplus/10.1108/09727980911007172.

Sarkis, Joseph (1999): How Green is the Supply Chain? Practice and Research. In SSRN Journal (SSRN Electronic Journal), PDF pp. 1-40. DOI: 10.2139/ssrn.956620. URL: http://www.researchgate.net/profile/Joseph_Sarkis/publication/228285308_How_Green_is_the_Supply_Chain_Practice_and_Research/links/0deec5177c79f8a809000000.pdf.

Schaltegger, Stefan and Burritt, Roger (2000a): Contemporary Environmental Accounting. Issues, Concepts and Practice. Sheffield: Greenleaf.

Schaltegger, Stefan; Hahn, Tobias and Burritt, Roger (2000b): Environmental Management Accounting - Overview and Main Approaches. Lueneburg: CSM. URL: http://www2.leuphana.de/umanagement/csm/content/nama/downloads/download_publikationen/06-1downloadversion.pdf.

Schaltegger, Stefan and Sturm, Andreas (1995): Öko-Effizienz durch Öko-Controlling. Zur praktischen Umsetzung von EMAS und ISO 14'001. Mit Anwendungsbeispielen der Unternehmen Flumroc AG, Mohndruck GmbH, Bank Sarasin & Cie. Stuttgart, Zürich: Schäffer-Poeschel; vdf, Hochschul-Verl. an der ETH Zürich.

Scherpereel, Cécile; van Koppen, C.S.A (Kris) and Heering, G.B.F (Frederique) (2001): Selecting Environmental Performance Indicators. The Case of Numico. In: GMI 33, pp. 97–114. URL: http://search.ebscohost.com/login.aspx?direct=true&profile=ehost&scope=site&authtype=crawler&jrnl=09669671&AN=7491402&h=r0QuMfs8AI3ZZxiBNUZmFa2beg6vHmpn4JKZ4AoHcnkmqO547Ql63gVvQ4qQpnz9pdP5k8ewtKey6jaAZQzlaQ%3D%3D&crl=f.

Schmidt-Bleek, Friedrich (2014): Grüne Lügen. Nichts für die Umwelt, alles fürs Geschäft - wie Politik und Wirtschaft die Welt zugrunde richten. München: Ludwig Verlag.

Schröder, Frank Roland; Holbach, Dirk and Müller-Kirschbaum, Thomas (2014): Henkel: Nachhaltigkeit in der Wertschöpfungskette - Von der Philosophie zur konkreten Umsetzung. In: D'heur, Michael (Ed.): CSR und Value Chain Management. Profitables Wachstum durch nachhaltig gemeinsame Wertschöpfung. Berlin, Heidelberg: Springer Gabler (SpringerLink : Bücher), pp. 257–268. URL: http://link.springer.com/content/pdf/ 10.1007%2F978-3-642-39889-6_11.pdf.

Schwarz, Jeanette; Beloff, Beth and Beaver, Earl (2002): Use Sustainability Metrics to Guide Decision-Making. Sustainability provides a framework for integrating environmental, social and economic interests into effective business strategies. Environmental Protection. In: CEP, pp. 58–63. URL: http://people.clarkson.edu/~wwilcox/Design/sustain.pdf.

Schwarz, Joshua L. and Murphy, Thomas E. (2007): Human Capital Metrics: An Approach To Teaching Using Data and Metrics To Design and Evaluate Management Practices. In: Journal of Management Education, Vol. 32 Iss. 2, pp. 164–182. URL: http://jme.sagepub.com/content/32/2/164.full.pdf.

Sedghi, Ami (URL): How much water is needed to produce food and how much do we waste? The Guardian. http://www.theguardian.com/news/datablog/2013/ jan/10/how-much-water-food-production-waste (20.06.2015).

Seichea, Catalin Florin (2009): Using Sustainable Development Indicators In The Quality Management Of Service Activities. Case Study: Veolia Environnement. In: Amfiteatru Economic, Vol. 11 Iss. 26, pp. 503–510. URL: http://www.amfiteatrueconomic.ro/temp/Article_891.pdf.

SERI (2009): Overconsumption? Our use of the World's Natural Resources. Friends of the Earth Europe, SERI, Global 2000. Sustainable Europe Research Institute (SERI). Vienna. URL: http://old.seri.at/documentupload/SERI%20PR/ overconsumption--2009.pdf.

Shaik, Mohammed and Abdul-Kader, Walid (2012): Performance measurement of reverse logistics enterprise: a comprehensive and integrated approach. In: Measuring Business Excellence, Vol. 16 Iss. 2, pp. 23–34. DOI: 10.1108/13683041211230294. URL: http://www.emeraldinsight.com/doi/ pdfplus/10.1108/13683041211230294.

Shen, Lixin et al. (2013): A fuzzy multi criteria approach for evaluating green supplier's performance in green supply chain with linguistic preferences. In: Resources, Conservation and Recycling, Vol. 74, pp. 170–179. DOI: 10.1016/j.resconrec.2012.09.006. URL: http://www.sciencedirect.com/ science/article/pii/S092134491200167X/pdfft?md5=f56b0479dd4279e8f2e03ce37e5a0ae7&pid=1-s2.0-S092134491200167X-main.pdf.

Shore, Sandy (URL): Wheat Prices Climb as Heat Takes Toll on Crops. Chicago Sun-Times. URL: http://finance.yahoo.com/news/wheat-prices-climb-heat-takes-195018096.html (22.04.2015).

Sikdar, Subhas K. (2003): Sustainable Development and Sustainability Metrics. In: AIChE J. (AIChE Journal), Vol. 49 Iss. 8, pp. 1928–1932. DOI: 10.1002/aic.690490802. URL: http://onlinelibrary.wiley.com/doi/10.1002/aic.690490802/pdf.

Singer, Alison (2013): State of the World: A Year in Review. In: Assadourian, Erik; Prugh, Tom; Starke, Linda (Eds.): Is Sustainability Still Possible? State of the World 2013. Washington, DC [etc.]: Island Press (State of the world, 2013), pp. xv–xxi.

Singh, Digvijay and Mathur, V. K. (2011): Greening of Supply Chain. In: Growth and Change, Vol. 39 Iss. 3, pp. 33–37. URL: http://www.sail.co.in/sites/default/files/publications/growth_dec2011.pdf.

Singhal, Pranshu et al. (2004): Key Environmental Performance Indicators (KEPIs): A New Approach to Environmental Assessment. LCA Forum. URL: http://www.lcaforum.ch/Portals/0/DF_Archive/DF27/Stutz2KEPIPaper2004.pdf.

Sirkin, Harold L. (URL): Sustainable Business Model Pays Off. Companies & Industries. BloombergBusiness. http://www.bloomberg.com/bw/articles/2013-11-26/a-sustainable-business-model-pays-off (03.04.2015).

Smith, John (2015a): Air Pollution: World's Worst Environmental Health Risk. In: UNEP (Ed.): UNEP Year Book 2014. Emerging Issues in Our Global Environment: United Nations Pubns, pp. 42–47. URL: http://www.unep.org/yearbook/2014/PDF/UNEP_YearBook_2014.pdf.

Smith, John (2015b): Excess Nitrogen in the Environment. In: UNEP (Ed.): UNEP Year Book 2014. Emerging Issues in Our Global Environment: United Nations Pubns, pp. 6–11. URL: http://www.unep.org/yearbook/2014/PDF/UNEP_YearBook_2014.pdf.

Smith, John (2015c): Securing Soil Carbon Benefits. In: UNEP (Ed.): UNEP Year Book 2014. Emerging Issues in Our Global Environment: United Nations Pubns, pp. 54–59. URL: http://www.unep.org/yearbook/2014/PDF/UNEP_YearBook_2014.pdf.

Sommer, Peggy (2010): Instrumente zur Unterstützung des Umweltmanagements. In: Kramer, Matthias (Ed.): Integratives Umweltmanagement. Systemorientierte Zusammenhänge zwischen Politik, Recht, Management und Technik. 1st ed. Wiesbaden: Gabler, pp. 323–383. URL: http://link.springer.com/content/pdf/10.1007%2F978-3-8349-8602-3.pdf.

Srivastava, Samir K. (2007): Green supply-chain management: A state-of-the-art literature review. In: Int J Management Reviews (International Journal of Management Reviews), Vol. 9 Iss. 1, pp. 53–80. DOI: 10.1111/j.1468-2370.2007.00202.x. URL: http://iic.wiki.fgv.br/file/view/Green%20supply-chain%20management.A%20state-ofthe-art%20literaturereview.2007.Srivastava..pdf.

Stubbs, Wendy and Cocklin, Chris (2008): An Ecological Modernist Interpretation of Sustainability: the Case of Interface Inc. In: Bus. Strat. Env. (Business Strategy and the Environment), Vol. 17 Iss. 8, pp. 512–523. DOI: 10.1002/bse.544. URL: http://www.researchgate.net/profile/Wendy_Stubbs/publication/227962959_An_ecological_modernist_interpretation_of_sustainability_the_case_of_Interface_Inc/links/00b7d53bca2ec82065000000.pdf.

Stubenrauch, Ina; Fach, Peter and Zuckarelli, Joachim (2014): Berichten Sie noch oder steuern Sie schon? Ansatzpunkte für ein pragmatisches Nachhaltigkeitscontrolling. In: Controller Magazin, Iss. 5, pp. 12–17.

The Economist (URL): Following the footprints. Carbon Footprints. Technology Quarterly: Q2 2011. http://www.economist.com/node/18750670 (10.07.2015).

The Greenhouse Gas Protocol (URL): All Tools. Cross Sector Tools. Calculation Tool - GHG emissions from transport or mobile sources. World Resources Institute; WBCSD. http://www.ghgprotocol.org/calculation-tools/all-tools (20.07.2015).

The Guardian (URL): What are CO2e and global warming potential (GWP)? Climate change - The ultimate climate change FAQ. http://www.theguardian.com/environment/2011/apr/27/co2e-global-warming-potential (05.06.2015).

The U Thant Institute (URL): Human Environment. http://www.uthantinstitute.org/index.php?option=com_content&view=article&id=130:human-environment&catid=67:human-environment&Itemid=126 (27.06.2015).

Tjahjadi, B. et al. (1999): Material and energy flow accounting in Germany—Data base for applying the national accounting matrix including environmental accounts concept. In: Structural Change and Economic Dynamics, Vol. 10 Iss. 1, pp. 73–97. DOI: 10.1016/S0954-349X(98)00060-5. URL: http://www.sciencedirect.com/science/article/pii/S0954349X98000605/pdfft?md5=b1f5f3c04e62ea28da20a17af59da8e2&pid=1-s2.0-S0954349X98000605-main.pdf.

Toppinen, Anne and Korhonen-Kurki, Kaisa (2013): Global Reporting Initiative and social impact in managing corporate responsibility: a case study of three multinationals in the forest industry. In: Bus Ethics Eur Rev (Business Ethics: A European Review), Vol. 22 Iss. 2, pp. 202–217. DOI: 10.1111/beer.12016. URL: http://web.a.ebscohost.com/ehost/viewarticle?data=dGJyMPPp44rp2%2fdV0%2bnjisfk5Ie46a9JrquzT7Ck63nn5Kx95uXxjL6srUq2pbBIr6ueSrims1KyrJ5Zy5zyit%2fk8Xnh6ueH7N%2fiVaunr1G0p7NJr6y0PurX7H%2b72%2bw%2b4ti7evLepIzf3btZzJzfhruutEmyq7dItZzkh%2fDj34y75uJ%2bxOvqhNLb9owA&hid=4106.

Trapp, Andrew C. and Sarkis, Joseph (2014): Identifying Robust portfolios of suppliers: a sustainability selection and development perspective. In: Journal of Cleaner Production, pp. 1–13. DOI: 10.1016/j.jclepro.2014.09.062. URL: http://www.sciencedirect.com/science/article/pii/S0959652614009925/pdfft?md5=c1b7b179ee3f8afff396666b17cb1077&pid=1-s2.0-S0959652614009925-main.pdf.

Tuzkaya, G. et al. (2009): Environmental performance evaluation of suppliers: A hybrid fuzzy multi criteria decision approach. In: International Journal of Environment Science and Technology, Vol. 6 Iss. 3, pp. 477–490. URL: http://link.springer.com/content/pdf/10.1007%2FBF03326087.pdf.

Tyteca, Daniel (1996): On the Measurement of the Environmental Performance of Firms— A Literature Review and a Productive Efficiency Perspective. In: Journal of Environmental Management, Vol. 46 Iss. 3, pp. 281–308. DOI: 10.1006/jema.1996.0022. URL: http://www.sciencedirect.com/science/article/pii/S0301479796900225/pdfft?md5=10123a3b6873c47fe31952e1b11ce991&pid=1-s2.0-S0301479796900225-main.pdf.

UCS (URL): Environmental Impacts of Renewable Energy Technologies. Union of Concerned Scientists. http://www.ucsusa.org/clean_energy/our-energy-choices/renewable-energy/environmental-impacts-of.html#.VYa3I0Ym_EY (10.07.2015).

UN Global Compact (2013): Global Corporate Sustainability Report 2013. United Nations Global Compact. URL: http://www.unglobalcompact.org/docs/about_the_gc/Global_Corporate_Sustainability_Report2013.pdf.

UNDP and UN DESA (2012): Synthesis Of National Reports For Rio+20. United Nations Department of Economic and Social Affairs (UN DESA), United Nations Development Program (UNDP), RIO+20. URL: https://sustainabledevelopment.un.org/content/documents/742RIO+20_Synthesis_Report_Final.pdf.

UNEP (2010): Green Economy. Developing Countries Success Stories. United Nations Environment Programme (UNEP). URL: http://www.unep.org/pdf/greeneconomy_successstories.pdf.

van Bergen, Barend (2010): Sustainable Insight. Your quarterly insight into sustainability. Integrated reporting; Closing the loop of strategy. KPMG Advisory N.V. (Ed.). KPMG. URL: https://www.kpmg.com/GR/en/IssuesAndInsights/ArticlesPublications/Sustainability/Documents/Sustainable-insight-April-2010.pdf.

Veleva, Vesela and Ellenbecker, Michael (2001): Indicators of sustainable production: framework and methodology. In: Journal of Cleaner Production, Vol. 9 Iss. 6, pp. 519–549. DOI: 10.1016/S0959-6526(01)00010-5. URL: http://www.sciencedirect.com/science/article/pii/S0959652601000105/pdfft?md5=cd13265f201c9b810751c0bca1660258&pid=1-s2.0-S0959652601000105-main.pdf.

Verfaillie, Hendrik A.; Bidwell, Robin and Cowe, Roger (2000): Measuring eco-efficiency. A guide to reporting company performance. Gemeva, Switzerland: World Business Council for Sustainable Development. URL: http://www.bcsd.org.tw/sites/default/files/node/domain_tool/110.file.128.pdf.

Viadiu, Frederic Marimon; Fa, Martí Casadesús and Saizarbitoria, Iñaki Heras (2006): ISO 9000 and ISO 14000 standards: an international diffusion model. In: Int Jrnl of Op & Prod Mnagemnt (International Journal of Operations & Production Management), Vol. 26 Iss. 2, pp. 141–165. DOI: 10.1108/01443570610641648. URL:http://www.emeraldinsight.com/doi/pdfplus/10.1108/01443570610641648.

Vijayvargy, Lokesh and Agarwal, Gopal (2014): Empirical Investigation of Green Supply Chain Management Practices and Their Impact on Organizational Performance. In: The IUP journal of supply chain management : IJSCM, Vol. 11 Iss. 4, pp. 25–42. URL: http://search.ebscohost.com/login.aspx?direct=true&profile=ehost&scope=site&authtype=crawler&jrnl=09729267&AN=100310810&h=bLnZH3aysOSUKlu0s7aNmKb5TPNntKRsS56zIcCZmj5F0XW4XwA0SIORhvJXAH6t7Zt6OPjbusU8MRkrkykv0A%3D%3D&crl=f.

Visser, Wayne (2008): Corporate Social Responsibility In Developing Countries. Chapter 21. In: Crane, Andrew Matten Dirk; McWilliams, Abagail; Moon, Jeremy; Siegel, Donald S. (Eds.): The Oxford handbook of corporate social responsibility. Oxford, New York: Oxford University Press (Oxford handbooks), pp. 473–499. URL: http://www.waynevisser.com/wp-content/uploads/2012/04/chapter_wvisser_csr_dev_countries.pdf.

Walley, Noah and Whitehead, Bradley (1994): It's Not Easy Being Green. In: Harvard business review on profiting from green business, Vol. 72 Iss. 3, pp. 46–51. URL: http://web.b.ebscohost.com/ehost/viewarticle?data=dGJyMPPp44rp2%2fdV0%2bnjisfk5Ie46a9JrquzT7Ck63nn5Kx95uXxjL6trUmvpbBIr6ueTLint1Kuqp5Zy5zyit%2fk8Xnh6ueH7N%2fiVbCmr0q2prFLt6qkhN%2fk5VXj5KR84LPgjOac8nnls79mpNfsVbeqrk6wqa9Os6ekfu3o63nys%2b585LzzhOrK45Dy&hid=118.

Wang, Xiaojun and Chan, Hing Kai (2013): A hierarchical fuzzy TOPSIS approach to assess improvement areas when implementing green supply chain initiatives. In: International Journal of Production Research, Vol. 51 Iss. 10, pp. 3117–3130. URL: http://web.b.ebscohost.com/ehost/viewarticle?data=dGJyMPPp44rp2%2fdV0%2bnjisfk5Ie46a9JrquzT7Ck63nn5Kx95uXxjL6trUmvpbBIr6ueTLint1Kuqp5Zy5zyit%2fk8Xnh6ueH7N%2fiVaursUi1qK9Mr6akhN%2fk5VXj5KR84LPgjOac8nnls79mpNfsVbastUy0p7ZIpNztiuvX8lXu2uRe8%2bLqbOPu8gAA&hid=118.

WBCSD and WRI (2004): The Greenhouse Gas Protocol. A Corporate Accounting and Reporting Standard. Rev. ed. Geneva, Switzerland, Washington, DC: World Business Council for Sustainable Development (WBCSD); World Resources Institute (WRI). URL: http://www.ghgprotocol.org/files/ghgp/public/ghg-protocol-revised.pdf.

WCED (1987): Report of the World Commission on Environment and Development: Our Common Future. URL: PDF pp. 1-300: http://www.google.de/url?sa=t&rct=j&q=&esrc=s&source=web&cd=1&sqi=2&ved=0CCYQFjAA&url=http%3A%2F%2Fwww.un-documents.net%2Four-common-future.pdf&ei=3xGhVbOLOof3Uv_ggTA&usg=AFQjCNHK0nnZ_enfeco-YZ_n0K7FwCx0VA&sig2=A1uaw3q6QvJjjIf1xO4CFA&bvm=bv.97653015,d.bGQ&cad=rja. Online Version : http://www.un-documents.net/wced-ocf.htm (31.03.2015). (Original Version: http://www.bne-portal.de/fileadmin/unesco/de/Downloads/Hintergrundmaterial_international/Brundtlandbericht.File.pdf?linklisted=2812).

Weber, Charles A.; Current, John R. and Benton, W.C (1991): Vendor selection criteria and methods. In: European Journal of Operational Research, Vol. 50 Iss. 1, pp. 2–18. DOI: 10.1016/0377-2217(91)90033-R. URL: http://www.sciencedirect.com/science/article/pii/037722179190033R/pdf?md5=0ed30baa5e7ebb99aa5bb643e0b737d7&pid=1-s2.0-037722179190033R-main.pdf.

Weber, Daniel and Bahr, Dorle (2014): Beiersdorf: Umsetzung gemeinsamer Wertschöpfung durch Kollaboration, Planung und Messung. In: D'heur, Michael (Ed.): CSR und Value Chain Management. Profitables Wachstum durch nachhaltig gemeinsame Wertschöpfung. Berlin, Heidelberg: Springer Gabler (SpringerLink : Bücher), pp. 123–137. URL: http://link.springer.com/content/pdf/10.1007/978-3-642-39889-6.pdf.

Weber, Jürgen (2015): Green Controlling - ein langer Atem ist erforderlich. In: Controller Magazin, Iss. 1, p. 17.

Weidema, Bo P. et al. (2008): Carbon Footprint. A Catalyst for Life Cycle Assessment? In: Journal of Industrial Ecology, Vol. 12 Iss. 1, pp. 3–6. URL: http://onlinelibrary.wiley.com/doi/10.1111/j.1530-9290.2008.00005.x/pdf.

Wiedmann, Thomas and Barrett, John (2010): A Review of the Ecological Footprint Indicator—Perceptions and Methods. In: Sustainability, Vol. 2 Iss. 6, pp. 1645–1693. URL: http://www.mdpi.com/2071-1050/2/6/1645/pdf.

Wiedmann, Thomas O. et al. (2013): The material footprint of nations. In: Proceedings of the National Academy of Sciences, pp. 6271–6276. URL: http://www.pnas.org/content/early/2013/08/28/1220362110.full.pdf.

Wiesehahn, Andreas (2015): Nachhaltige Wirtschaftlichkeitsbeurteilung. In: Controller Magazin, Iss. 3, p. 29.

Wilson, Jeffrey; Tyedmers, Peter and Pelot, Ronald (2007): Contrasting and comparing sustainable development indicator metrics. In: Ecological Indicators, Vol. 7 Iss. 2, pp. 299–314. URL: http://www.researchgate.net/profile/Peter_Tyedmers/publication/223804576_Contrasting_and_comparing_sustainable_development_indicator_metrics/links/0912f4ffee3cf25ee6000000.pdf.

Wolters, Teun and Danse, Myrtille (2002): Towards Sustainability Indicators for Product Chains. With Special Reference to an International Coffee Chain. In: Bennett, Martin; Bouma, Jan Jaap; Wolters, Teun (Eds.): Environmental Management Accounting: Informational and Institutional Developments. Dordrecht [etc.]: Kluwer Academic Publishers (Eco-Efficiency in Industry and Science, 9), pp. 231–247.

Writers, Staff (URL): Snow blocks in tens of thousands as cold death toll rises. Terra Daily. Belgrade (AFP). http://www.terradaily.com/reports/Snow_blocks_in_tens_of_thousands_as_cold_death_toll_rises_999.html (22.04.2015).

Yang, Ching-Chow; Yang, King-Jang and Peng, Shu-Yun (2011): Exploration strategies and key activities for the system of environmental management. In: Total Quality Management & Business Excellence, Vol. 22 Iss. 11, pp. 1179–1194. DOI: 10.1080/14783363.2011.603201. URL: http://www.tandfonline.com/doi/pdf/10.1080/14783363.2011.603201.

Zirkler, Bernd; Zimmermann, Frank and Theile, Markus (2011): Standardisierung von Kennzahlen im Green Controlling. Horváth & Partners Management Consultants. In: rechnungswesen & controlling, pp. 2–3. URL: http://www.horvath-partners.com/fileadmin/horvath-partners.com/assets/05_Publikationen/PDFs/deutsch/E_r_c_4-2011_Kennzahlen_im_Green_Controlling_Zirkler-Zimmermann-Theile.pdf.

Alphabetical Index

U

V

W

Monograph Series MEYER STIFTUNG

Edited by Prof. Dr. Claus Meyer

Volume 1: Sybille Molzahn, Die Bilanzierung der betrieblichen Altersversorgung nach HGB und IFRS, 2., überarb. u. erw. Aufl. 2007, ISBN 978-3-89673-432-7

Volume 2: Paul Pronobis, Das Umsatzkostenverfahren im internationalen Vergleich. Beschreibung des Aufbaus sowie der einzelnen Posten nach HGB, IFRS und US-GAAP, 2007, ISBN 978-3-89673-425-9

Volume 3: Veronika Trauth, Sukzessive Unternehmenserwerbe/-veräußerungen im Konzernabschluss nach IFRS. Darstellung, Würdigung, Beispiele, 2007, ISBN 978-3-89673-433-4

Volume 4: Patrick Krauß, Publizität von Abschlussprüferhonoraren bei kapitalmarktorientierten Unternehmen. Zielsetzung und Wirkung der Regelungen im Bilanzrechtsreformgesetz, 2008, ISBN 978-3-89673-446-4

Volume 5: Jürgen Halter, Werthaltigkeitsprüfung von zahlungsmittelgenerierenden Einheiten nach IAS 36. Darstellung und konzeptionelle Kritik unter besonderer Berücksichtigung des Nutzungswerts, 2008, ISBN 978-3-89673-468-6

Volume 6: Carolin Schwarz, Kaufpreisvereinbarungen im Rahmen von Unternehmensakquisitionen und deren bilanzielle Behandlung nach IFRS, 2008, ISBN 978-3-89673-490-7

Volume 7: Friederike Maier, Rückstellungen nach IFRS. Kritische Analyse und aktuelle Entwicklungen unter besonderer Beachtung von Entsorgungs- und Wiederherstellungsverpflichtungen, 2009, ISBN 978-3-89673-515-7

Volume 8: Barbara Stütz, Steuerwettbewerb in Europa, 2009, ISBN 978-3-89673-530-0

Volume 9: Viktoria Zerr, Ansatzpunkte zur Optimierung des Controllingsystems im kommunalen Immobilienmanagement, 2010, ISBN 978-3-89673-551-5

Volume 10: Christian Friedel, Die ertragsteuerliche Behandlung und deren Gestaltungsmöglichkeiten bei der GmbH & atypisch Still, 2., überarb. u. erw. Aufl. 2013, ISBN 978-3-89673-648-2

Volume 11: Fatma Ünal, IFRS 11 Joint Arrangements: Darstellung, kritische Würdigung anhand von comment letters & Analyse der Auswirkungen auf die Unternehmensbonität, 2013, ISBN 978-3-89673-660-4

Volume 12: Christian Kern, Ausgewählte Änderungen und Auswirkungen von IAS 19 (revised 2011), 2014, ISBN 978-3-89673-672-7

Volume 13: Viktoria Göbel, Corporate Intellectual Capital Reporting: the Case of Germany, 2014, ISBN 978-3-89673-677-2

Volume 14: Caroline Hauber, Die ertragsteuerliche Behandlung des Nießbrauchs an Personengesellschaftsanteilen, 2015, ISBN 978-3-89673-686-4

Volume 15: Kevin Blum, Die Anwendung des § 6 Außensteuergesetz (AStG) zur Wegzugsbesteuerung natürlicher Personen, 2015, ISBN 978-3-89673-697-0

Volume 16: Claus Meyer, MEYER-STIFTUNG – Ein Bericht über die Jahre 2005 bis 2015, 2015, ISBN 978-3-89673-706-9

Volume 17: Andreas Boll, Goodwill-Bilanzierung im Konzernabschluss kapitalmarktorientierter Unternehmen, 2016, ISBN 978-3-89673-721-2

Volume 18: Tanja Schillinger, An accounting approach to create an environmentally sustainable company: Selection and Definition of Environmental Indicators with special reference to Suppliers in Developing Countries, 2016, ISBN 978-3-89673-722-9

Patrons

BLANCO

Rotary Club
Pforzheim-Schloßberg

Promoters

MICROTHERM

MIC
Microtherm International Cooperation

Sparkasse
Pforzheim Calw